THE AMERICAN FARM CRISIS

An Annotated Bibliography with Analytical Introductions

RESOURCES ON CONTEMPORARY ISSUES

RESOURCES ON CONTEMPORARY ISSUES

Richard A. Gray
Series Editor

Entirely devoted to topics that address important social, economic, and political concerns, RESOURCES ON CONTEMPORARY ISSUES is a new bibliographic series that guides the reader to the most significant literature available on these subjects. Every book in this series is designed for use in school, academic, and public libraries. Librarians, teachers and students will find them useful not only as a means of organizing the literature of the topic, but also as a framework for understanding the complexity of the issue.

RESOURCES ON CONTEMPORARY ISSUES is one of the few bibliographic series to consistently provide analytical introductions--background text which explains the significance of the topic and provides a context in which to understand the citations assigned to each chapter. Each volume contains 800 to 1,000 annotated citations, a glossary, a chronology of events, and author and title indexes.

Other volumes in this series:

Endangered Species
Nicaragua
Protectionism
South Africa
Terrorism

This Series is Available on Standing Order.

THE AMERICAN FARM CRISIS

An Annotated Bibliography with Analytical Introductions

by

Harold D. Guither
Professor of Agricultural Economics
University of Illinois at Urbana-Champaign

and

Harold G. Halcrow
Professor Emeritus
Department of Agricultural Economics
University of Illinois at Urbana-Champaign

RESOURCES ON CONTEMPORARY ISSUES

Pierian Press
Ann Arbor, Michigan

1988

The Pierian Press
P. O. Box 1808
Ann Arbor, Michigan 48106

Printed and Bound in the United States of America

Library of Congress Cataloging-in-Publication Data

Guither, Harold D.
The American Farm Crisis

Resources On Contemporary Issues
Includes indexes.
1. Agriculture--Economic aspects--United States--Bibliography. 2. Agriculture and state--United States--Bibliography. 3. Farms--United States--Bibliography. 4. Farms, Small--United States--Bibliography.
I. Halcrow, Harold G. II. Title. III. Series.
Z5075.U5G85 1988 [HD1761] 016.3381'0973 87-32846
ISBN 0-87650-240-0

Table of Contents

Table of Contents v

Preface ix

Introduction xi

Chapter 1: The American Farm in Transition 1

To understand the complex issues surrounding the American farm requires a broad understanding of its basic features, the size and income distribution of farms, the diversity of crop and livestock production, and the relationship of government in the food and fiber system.

The Changing Farm Structure 1

Diversity and Specialization Among Farms 3

Farm Financial Problems 5

Annotated Bibliography 9

Chapter 2: The Scientific and Technological Revolution in Farming 15

Technology is the key to productivity of the American farm and our competitive position in the world. A myriad of new technological advances will dominate growth prospects for the American farm over the next generation. How these technologies are developed, used, or managed is a crucial issue.

Technology and Productivity 15

The New Age of High Technology 17

Human Capital, Welfare, and Environment 19

Annotated Bibliography 20

Chapter 3: The Evolution in Farm Business Management 27

Revolutionary changes in the financial, economic, and technological environment in which farmers have influenced and accompanied a revolution in farm business management. Prices are more volatile, risks are higher, capital gains and losses have occurred. Cash flow and tax management are part of the many concerns facing farmers.

Types of Farm Business Organization 28

The Beginnings of Farm Management as a Science 28

Farmer Participation and Involvement 29

Professional Farm Management Services 29

Credit and Financial Analysis in the Farm Business 30

Tax Management and Farm Management 30

The Computer Revolutionizes Farm Business Management 31

Causes and Effects of Advanced Management Technology 31

Annotated Bibliography 32

Chapter 4: The Evolution in Markets and Marketing of Agricultural Products 41

A principal problem in agriculture has been the difficulty of coordinating production with market needs. Lack of change in the way individual farmers market their products is often blamed for oversupply, lower prices, and total failure of a farming operation.

Chapter 5: The Farm and Its Setting in the Rural Community 59

The connection between the American farm and the surrounding community has changed profoundly. The farm input market has become an important part of the business of the farming community. Many rural communities have also developed nonfarming industries and many farmers and their families beneift from employment off the farm.

Chapter 6: Government Farm Commodity Programs 73

Current government farm programs aim to maintain farm income with emphasis on family farms, keep major export commodities competitive in world markets, and limit governmental budget costs within limits that are politically acceptable. Continuing present policies will become increasingly expensive.

Chapter 7: Transforming Traditional Goals for Times Ahead 95

The financial crisis of the American farm was created by the mix of past and current policy. It will intensify until something fundamental is changed. Major changes should include reduction of government costs, improving quality and quantity of exports, and refinancing and restructuring of the farm debt.

Chapter 8: Educational Programs for Implementing Change 121

The modernization of farming may be partly credited to investments in education, skills and health of people in farming, agribusiness, and industry related to agriculture. Farmers and farming operations will continue to face adjustments perhaps even more revolutionary than in the past.

Chronology

Glossary

Indexes

Preface:

RESOURCES ON CONTEMPORARY ISSUES

Scope

The topics chosen for this series are those controversial issues whose resolution presupposes an intervention of government. The crisis on the farm amply illustrates the point. While some may argue that the crisis itself is primarily the creation of past governmental interventions, nonetheless any resolution of the present crisis requires that the government intervene in other ways that differ, markedly perhaps, from those of the recent past. Future volumes in the series will be devoted to topics that likewise pivot on governmental policy and decision making.

Purpose

Resources on Contemporary Issues (RCI) volumes are designed to serve as gateways or points of entrance to the bibliography of current social issues. Directed to trained scholars as well as to non-specialist readers, they provide contextual, analytical guidance to complex, inter-related bodies of literature.

Structure

Each chapter in an RCI volume begins with a thorough, analytical introduction that is divided into sub-sections. These sub-sections are then echoed in the classified bibliography for that chapter. The citations that constitute any discrete segment of the classified bibliography are, by this means, firmly anchored in the corresponding section of the analytical introduction. The annotations that follow each citation likewise have a contextual structure. Each has a valuative element, even if only implied; a full descriptive or indicative element; and a substantive element that may be either a paraphrase or a direct quotation from the book or article under consideration.

Format

The information published in individual volumes of this series has been produced, and is maintained, on computers. The information is available in print and in electronic formats.

Introduction:

THE AMERICAN FARM CRISIS

Harold D. Guither and **Harold G. Halcrow**

The nation with too much bread has many problems; the nation with too little bread has only one problem.
Byzantine Proverb (5th Century, A.D.)

Approaching the 1990s, a higher percentage of American farms are under economic stress than at any time since the Great Depression of the 1930s. Adverse economic conditions are causing a sharp decline in the number of family farms, especially the smaller ones. Some farmers of all sizes are having financial problems. Government programs, designed to help the American farmer, have become so expensive that they are in jeopardy. Efforts to reduce program costs have become an urgent priority.

Yet today, the American farm stands at a frontier; a large array of new, high technology is becoming available to greatly increase production. By 2010, with continuing support for research and development, the total output of American farms can increase by 60 percent, and by 2030 output can be double that of 1980. Abundant production will make it possible to export a greater amount of total farm output than can be used in the domestic market.

This greater output may be sorely needed. World population is expected to increase from five billion people in 1985 to six billion by 1995 or 1996. At projected rates of growth, world population could reach nearly eight billion by 2010, and ten to eleven billion by 2030. According to current prospects, as much as five-sixths of the total increase will occur in the developing countries, and one-sixth in the developed countries of North America, Western Europe, Japan, Australia and New Zealand.

Over the last few years, except for Africa, the developing countries have been increasing their food production more rapidly than their population, but as economic development has occurred, they also have increased their imports of food from the developed countries. Further increases in exports of food from the developed to the developing countries is a rational expectation for the future, providing that economic development is successful and trade is not restricted by governmental policies of either.

This book deals with the advances in agriculture from an historic perspective, emphasizing the question of how best to manage the current situation. The authors are optimistic that the American farm faces a bright future. A comprehensive view of the current problem requires an examination of the status of farm people, the scientific and technological revolution sweeping the American farm economy, farm production potentials, domestic and world markets, and the necessary conditions for American farms to grow and prosper. The book discusses some of the necessary adjustments that must occur in the farm economy, the steps by which these adjustments may be made, and the broader effects on farm people and the rural, national, and world economy.

This first volume in the *Resources on Contemporary Issues* series serves as a gateway to the bibliography of the farm crisis in the United States, in all of its complex ramifications. Each of the eight chapters in the book opens with a long, analytical introduction that is divided into sub-sections. These sub-sections are in turn reflected in the classified bibliography for that chapter. All citations are thoroughly annotated, providing valuative, descriptive, and substantive elements. The work contains a glossary of terms and a chronology. Author and title indexes complete the work.

CHAPTER 1

THE AMERICAN FARM IN TRANSITION

"It was the best of times, it was the worst of times, it was the age of wisdom, it was the age of foolishness...."

Charles Dickens (1859), *A Tale of Two Cities*.[1]

To understand the complex issues surrounding the American farm requires a broad understanding of its basic features, the size and income distribution of farms, the diversity of crop and livestock production and the relationship of government in the food and fiber system.

The American farm is in a stage of rapid transition to a system of larger but fewer operating units. This chapter shows the size distribution of farms according to sales of farm commodities, gross cash farm income, direct government payments received per farm, off-farm income, and total net income. It projects trends in farm size, discusses farm diversity and specialization, and describes some of the farmers' financial problems.

Later chapters will deal with the technological revolution in farming, and provide a more comprehensive overview of the historical development, institutional structure, and future social and economic choices for the American farm, including suggested roles of government in respect to farming.

The Changing Farm Structure

Farm structure deals with the organization and ownership of resources for producing food and fiber, mostly for sale. A farm is defined by the United States Department of Agriculture and the United States Bureau of the Census as any operation with actual or potential sales of commodities of $1,000 or more in a single year. This permits some very small farming enterprises to be classed as farms. For example, commodities that are worth $1,000 can be obtained from as little as one good dairy cow, five acres of corn, or one-half acre of tobacco. There is no upward limit.

So farms range from small enterprises, which are of little commercial significance, to very large farms with annual gross farm sales of $500,000 and up. Within the full range of the American farms, gross cash farm income is highly skewed. For instance, as shown in Table 1.1, in 1985, some

Table 1-1.

Distribution of Farms by Value of Farm Sales and Gross Cash Farm Income, 1985

Farm size and sales class	No. of farms (thousands)	Percent of all farms	Gross cash farm income: billions	Percent of gross cash farm income
Very large - $500,000 and up	27	1.2 percent	$ 50	32.2 percent
Large family - $250,000-$499,999	66	2.9	26	16.6
Medium family - $40,000-$249,999	544	13.9	63	40.9
Small family - $10,000-$39,999	473	20.8	12	7.5
Rural residence less than $10,000	1,164	51.2	4	2.9
All farms	2,275	100.0	$156	100.0

Source: United States Department of Agriculture, Economic Research Service. *Economic Indicators of the Farm Sector, National Financial Summary, 1985.* (Washington, DC, November 1986): 42, 45. Percentages may not add to 100 due to rounding.

27,000 farms, which were only 1.2 percent of all farms, received 32.2 percent of the gross cash farm income. At the other end of the scale, more than one million small rural residence farms were 51.2 percent of all farms, and received only 2.9 percent of the gross cash farm income. In between these extremes, there were more than one million family farms, large, medium, and small, which totaled 47.6 percent of all farms and received 65.0 percent of the gross cash farm income.

The percentages in Table 1.1 are not stable. The total number of farms has been declining steadily for more than half a century. Farm numbers peaked at 6.8 million in 1935 and declined to 2.275 million in 1985. The U.S. Congress' Office of Technology Assessment has projected 1.8 million farms by 1990 and 1.25 million by 2000.[2]

The very large and the large family farms are increasing their share of the market while the medium and small family farms, and the rural residence farms are declining.

During some periods in the past, some of the relative growth was due to inflation in prices of farm products, but even when the effects of inflation are removed (sales measured in 1982 dollars), the growth of large farms is substantial. For instance, between 1974 and 1982 the very large farms increased their share of the market from a little more than one-fourth (25.2 percent) to almost one-third (32.5 percent); and, proportionally speaking, the large family farms grew even more (from 11.5 to 15.1 percent). The other three classes all declined in terms of their share of the market. Between 1982 and 1985, the medium family farms and small family farms continued to decline relative to the other classes. This trend is projected to continue through the 1980s and 1990s.

The average income per farm operator, or operator family, is composed of the net farm income, the value of direct government payments, and the operator family off-farm income. Both net farm income and the value of government payments, which are largely determined by the value of commodities covered under government programs, are highly skewed. For instance, the 27,000 very large farms had average net farm incomes of $640,000 in 1985, while the rural residence farms with sales

Table 1-2.

Trends in Distribution of Farm Sales by Size Groups.

Farm size and sales class	1974[1]	1978[1]	1982[1]	1985[2]
		(Percent)		
Very large-$500,000 and up	25.2	29.7	32.5	32.2
Large family-$250,000-499,999	11.5	13.2	15.1	16.6
Medium family-$40,000-249,999	46.0	43.6	41.5	40.9
Small family-$10,000-39,999	13.6	10.4	8.2	7.5
Rural residence-less than $10,000	3.9	3.1	2.7	2.9
All farms	100.0	100.0	100.0	100.0

1. In constant 1982 dollars; excludes abnormal farms.
2. In 1985 dollars.

Source: *Census of Agriculture*, 1974, 1978 and 1982. (Washington, DC, United States Department of Commerce, Bureau of the Census). United States Department of Agriculture Economic Research Service, *Economic Indicators of the Farm Sector, National Financial Summary, 1985.* (Washington, DC, November 1986): 45.

of less than $5,000 had net farm losses of $4,288. Direct payments averaged $37,499 for the very large farms and only $40 for the smallest rural residence farms. Off-farm income is more evenly distributed, generally increasing as size of farm decreases.

Although the rural residence farms show net farm losses, their real income is often higher than the data show. Some of these farms raise much of the food used by the farm family. Others get tax benefits by deducting farm losses against off-farm income. Great variations exist among rural residence farms in respect to incomes of the people who live on them, their reasons for living on a farm, and their degree of satisfaction with their living conditions.

Diversity and Specialization Among Farms

In addition to the wide range in size of farms, there is wide variation in diversity and specialization among farms. For instance, farms that get the major portion (more than 50 percent) of their income from crops are called crop farms, and these vary widely in diversity and specialization, as well as in size and income.[4] Most of the commercial crop farms in the United States grow more than one crop. The degree to which a farm specializes in the production of its principal crop increases with farm size, but some types of farms are more specialized than others. Corn is grown on more farms than any other crop in the United States and is the most significant crop in terms of value. However, corn farmers are less specialized than other crop farms as measured by the ratio of principal crop sales to total farm sales because many corn farmers have livestock and dairy enterprises and feed their corn to animals. About one-third of annual corn production is fed to livestock and poultry on farms where it is raised.

Soybeans, the second most valuable crop in the United States, are the leading crop in about one-quarter of the commercial crop farms. Soybeans are also grown on other farms where they are rotated with corn or sometimes double-cropped with wheat. A small percentage of soybeans are planted after winter wheat and harvested the same year. Nearly 60 percent of soybean farms are in Illinois,

Table 1-3.

Average per Operator Net Farm Income, Direct Government Payments, Off-Farm Income, and Total Income; Farms Classed by Value of Farm Sales, 1985.

Farm size and sales class	Net farm income	Direct government payments	Off-Farm income	Total income
Very large-$500,000 and up	$640,010	$ 37,499	$15,448	$692,957
Large family-$250,000-499,999	99,661	21,783	11,447	132,891
Medium family-$100,000-249,999	36,660	12,845	10,551	60,056
Medium family-$40,000-99,999	6,566	5,193	10,347	22,106
Small family-$20,000-39,999	-48	2,040	14,333	16,325
Small family-$10,000-19,999	-1,978	678	18,916	17,616
Rural residence $5,000-9,999	-1,211	233	21,538	20,557
Rural residence Less than $5,000	-4,288	40	22,644	18,496
All farms	13,881	3,387	17,945	35,213

Source: United States Department of Agriculture, Economic Research Service. *Economic Indicators of the Farm Sector, National Financial Summary, 1985.* (Washington, DC, November 1986): 49,51,52.

Iowa, Missouri, Ohio, and Indiana. Other states with relatively large numbers of soybean farms are North Carolina, Arkansas, Georgia, Tennessee, and Mississippi.

Wheat is third in total value among field crops. Production is dominated by very large farms and large family farms that grow wheat as their principal crop. Most wheat farms are in the Great Plains states from Texas to North Dakota, including Colorado and Montana, and a tenth are in the Northwest. Specialization in wheat farming increases with size, as many small farms that grow wheat are predominantly livestock and dairy farms.

Farms that derive their major portion of income from livestock vary just as widely in diversity and specialization as crop farms. For instance, beef cattle are raised on small family farms as well as on very large farms and ranches, but the clear trend has been for both cow-calf operations and cattle feeding enterprises to grow in size and become more specialized, as automation and other technology give advantages to large scale. Dairy production tends to be concentrated in large, highly automated farms. The smaller, less automated dairy farm has almost disappeared. Hog production is tending to be concentrated in more highly automated large family and very large farms. On many farms, livestock production is no longer a relatively

Figure 1.1.

Commercial Farms by Type and Financial Condition, 1985.

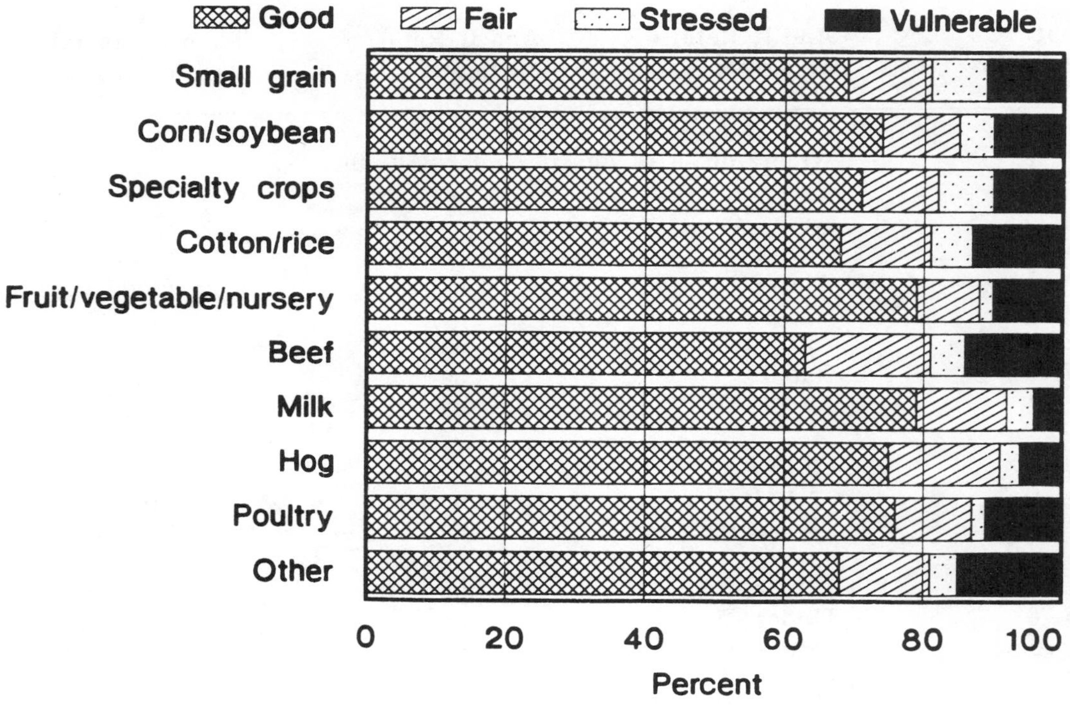

Source: *Outlook '87 Charts*, 63rd Annual Agricultural Outlook Conference, December 2-4, 1986. United States Department of Agriculture, Economic Research Service. Washington, DC: 1986, p. 91.

Note: The definition of "fair," "stressed," and "vulnerable" is provided in Table 1-4.

small supplementary enterprise, as it was traditionally, but a highly automated, capital-intensive operation.

Farm Financial Problems

Because of the wide diversity in enterprises, size of farm, and other economic factors, it is difficult to draw general conclusions about the financial status or problems of farm people. Some individuals who own farms live in towns and cities. If their farms are rented to tenants, they are not counted as farmers, but as landlords. The official U.S statistics call the tenant the farmer, not the landlord. It is common, however, for a tenant to rent from more than one landlord, sometimes from four or five, or more. Most of the very large and many of the large family farms have several landlords.

Most landowners and a high percentage of farm families are financially solvent. Some are rich, but in 1985, according to estimates proposed by the staff of the U.S. Department of Agriculture, the financial condition of nearly 25 percent of all commercial farms, not including those rural residence farms with gross sales of less than $5,000, were in some kind of financial difficulty, with debts that might be too high for them to manage compared with the income available for servicing debt. As many as one-third of the 25 percent were classed as "vulnerable," which may imply that they could not survive as economic units without substantial help. Farms in a second group were classed as "stressed." With luck and some help they might survive. The final third were classed as "fair," suggesting that their combination of debts and earnings was such that they could survive only a limited time without an increase in income or other help.

Table 1-4

Criteria for Commercial Farms Classification

If Debt/ Asset Ratio	And if Return on Assets	And if Return on Equity	Then Financial Position is
	Operators with Equity Under $50,000		
Under 40	Above 0	na	Good
40 to 70	Above 5	na	Good
Over 70	Above 15	na	Good
Under 40	-5 to 0	na	Fair
40 to 70	0 to 15	na	Fair
Over 70	5 to 15	na	Fair
Under 40	-5 to 5	na	Stressed
40 to 70	-5 to 0	na	Stressed
Over 70	0 to 5	na	Stressed
Under 40	Under -15	na	Vulnerable
40 to 70	Under -5	na	Vulnerable
Over 70	Under 0	na	Vulnerable
	Operators with Equity Above $50,000		
Under 40	Above 0	Above 0	Good
40 to 70	Above 5	Above 5	Good
Over 70	Above 15	Above 15	Good
If Not Already Classified as Good, Then			
Under 10	Above -15	Above -15	Fair
10 to 40	Above -5	Above -5	Fair
40 to 70	Above 0	Above 0	Fair
Over 70	Above 5	Above 5	Fair
If Not Already Classifed as Good or Fair, Then			
Under 10	na	na	Stressed
10 to 40	Above -15	Above -15	Stressed
40 to 70	Above -5	Above -5	Stressed
Over 70	Above 0	Above 0	Stressed
If Not Already Classifed as Good, Fair, or Stressed, Then Vulnerable			

na = not applied

Source: Based on a paper by Emanuel Melichar, "The Farm Credit Situation and the Status of Agricultural Banks," read at the Twin Cities Agricultural Issues Round Table, St. Paul, MN, 24 April 1986.

Figure 1.2.

Farms Selling $40,000+ by Type and Financial Condition, 1985.

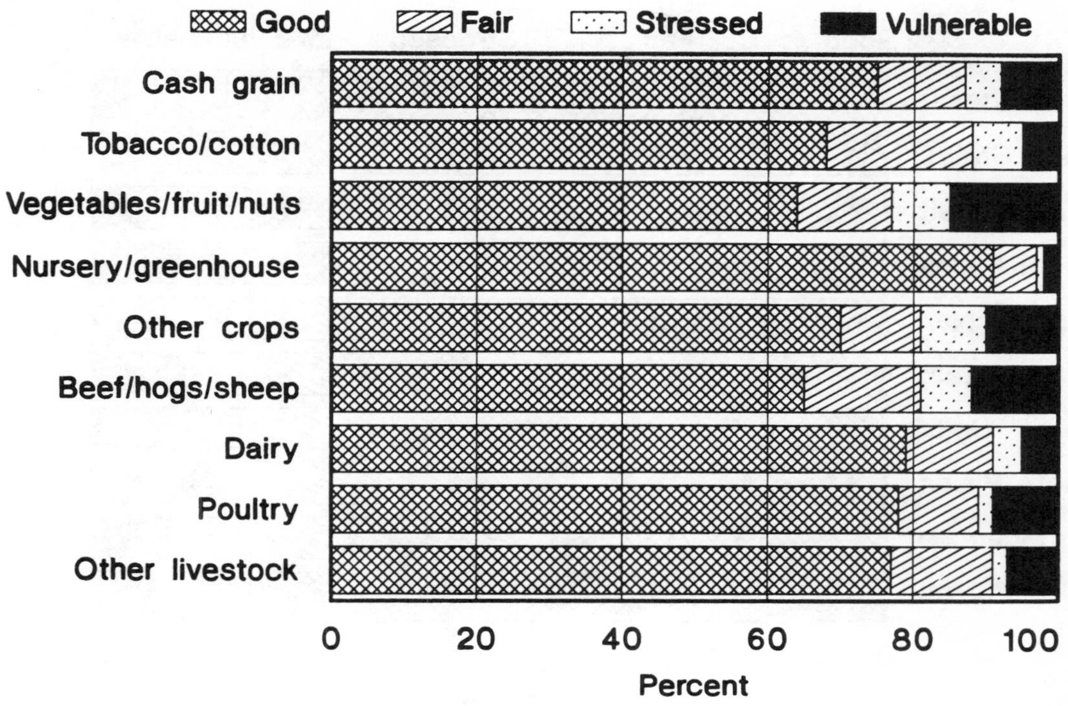

Source: Same as Figure 1.1, p. 93.

A large percentage of the farms that are in fair condition, stressed, or vulnerable, will not survive as economic units unless farm incomes improve or help is targeted to their needs. Some others of course will also fail, or be reorganized into more efficient enterprises. In 1985, more than 200,000 farmers out of 1.7 million in a survey by the U.S. Department of Agriculture were in financial stress or vulnerable due to a high debt load and an inability to generate enough cash to pay their bills. The federally-sponsored farmer-owned Farm Credit System reported a record $368 million in loan losses in 1984, with heavier losses expected in the years to follow. Nearly one-third of all farms with annual sales over $40,000 were facing some financial difficulties. This would force many farmers to discontinue their own operations with losses to farm lenders and readjustments in the structure of farm debt.

Typically, the land in these farms would be integrated with the resources of other operating farms, the farm family would retire or seek other work, and production in the farm sector would continue under other management. Traditionally, the farm population has declined as a result of reduced entry of new young farm families, with only a small percentage of the decline due to early retirement. Within a decade, as many as 400,000 farm families will be forced, as a result of financial difficulties, to discontinue as farm operators; and this is a major root of the current farm problem.

Fortunately, there is another, brighter side to the farm picture. Total farm debt (including operator households and loans from the Commodity Credit Corporation), which peaked at over $220 billion in the summer of 1983 after having risen each year since 1945, dropped to about $188 billion at the end of 1986.[5] Debt excluding CCC loans, the relevant debt for studies of financial stress, which peaked at $205 billion (on a year-end basis) in 1983, declined by $36 billion to about $169 billion at the end of 1986. Debt secured by farm real estate had declined more slowly than the non-real estate debt.

The ongoing reductions in farm debt can be attributed in large part to the following: (1) Lenders were charging off substantial amounts of debt

Figure 1.3.

Farms Selling $40,000+ by Region and Financial Condition, 1985.

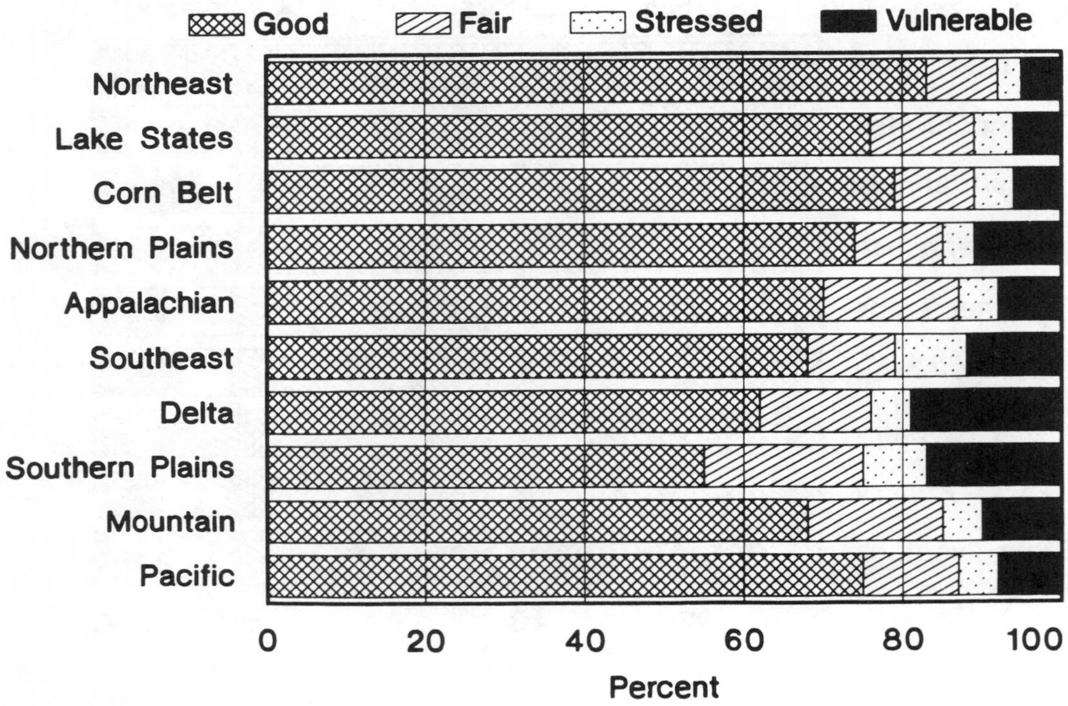

Source: Same as Figure 1.1, p. 95.

(commercial banks charged off $3.4 billion in non-real estate loans in the three years up to the end of 1986). (2) Farm assets were being transferred from heavily indebted owners to cash holders. (3) Borrowers with liquid assets, such as bank deposits and bonds, were cashing these to pay off debts. (4) Demand declined for new loans because of crop-acreage cutbacks, lower fuel prices, and reduced inventories of machinery and livestock or other farm assets. Banks increased their farm real estate loans by more than 10 percent annually over the four years 1983-86, while all other lenders except for the Farmers Home Administration decreased their holdings.

Adding to this picture, it is most important to recognize that markets for farm commodities will increase substantially over the next two or three decades, with the greatest possible growth coming from population increases around the world and increases in purchasing power by people in developing countries. The current challenge for policy is to help in developing these markets while implementing the areas of growth that are possible for the American farm. Much of the U.S. farm economy is in a stage where rapid growth is possible and almost certain to occur if the right choices are made in public policy.

In the next chapter we will examine some of these growth prospects and set the stage for a broader and more penetrating study of the American farm economy.

Endnotes

1. Charles Dickens. *A Tale of Two Cities*. London: 1859.

2. United States. Congress. Office of Technology Assessment. *Technology, Public Policy, and the Changing Structure of American Agriculture*. OTA-F-185: 96. Washington, DC: U.S. Government Printing Office, March 1986. This estimate is based on a Markov chain projection using a 1969-82 base, page 96.

3. Donn A. Reimund, Nora L. Brooks, and Paul D. Velde. *The U.S. Farm Sector in the Mid-1980s.*

Agricultural Economics Report, no. 548. U.S. Department of Agriculture, Economic Research Service, May 1986.

4. United States. Congress. Congressional Budget Office. *Diversity in Crop Farming: Its Meaning for Income Support Policy.* Special Study. Washington, DC, May 1985.

5. Emanuel Melichar. *Farm Credit Developments and the Financial Condition of Agricultural Banks: A Preliminary Report for the National Agricultural Credit Committee, March 16, 1987.* Washington, DC: Board of Governors of the Federal Reserve System.

Chapter 1: Annotated Bibliography

The Changing Farm Structure

*** 1.1 ***
Brewster, David E., Wayne D. Rasmussen, Garth Youngberg, eds. *Farms in Transition, Interdisciplinary Perspectives on Farm Structure.* Ames: Iowa State University Press, 1983. ISBN 0-8138-0636-4.

The authors define farm structure as "the way in which agriculture's resources are organized and controlled." Their book is a relatively broad study of the subject. Their basic theme is that "Farm structure, however defined, cannot be understood, studied, or treated except as an integral part of a larger mix of economic, political, and social forces sweeping us toward some unknown but certainly different future. The recent national dialogue on farm structure contributed in an important way to what is becoming a far-ranging and fundamental re-evaluation of this country's agricultural, food, and rural policies and goals."

*** 1.2 ***
Harrington, David, and Thomas A. Carlin. *The U.S. Farm Sector: How Is It Weathering the 1980s?* Agriculture Information Bulletin, no. 506. Washington, DC: Economic Research Service, U.S. Department of Agriculture, 1987.

The authors' study is based on the ninth report to Congress on family farms, submitted in accordance with the Food and Agriculture Act of 1977 (section 102), the Agriculture and Food Act of 1981 (section 1608), and the Food Security Act of 1985 (section 1441). In general summary, "Commercial farms with gross annual sales of $40,000 or more, 28 percent of all farms, generally had positive after-tax rates of return to equity in 1985. But noncommercial farms, those with gross annual sales of less than $40,000, 72 percent of all farms, showed small after-tax losses. The farm economy has deteriorated since 1981 when farm land values began to decline. By 1984, farming households earned only about 80 percent as much as the national average, compared with their historic high in 1973 when they earned almost 50 percent more than the national average. As many as 15 percent of all farm operators who were in business before 1980 may leave farming for financial reasons before the current economic adjustments end. Rural counties and communities whose economies rely on agriculture will have trouble maintaining many services as declining farm land values shrink tax revenues."

*** 1.3 ***
Harrington, David H., and Alden C. Manchester. "Profile of the U.S. Farm Sector." *Agricultural-Food Policy Review: Commodity Program Perspectives.* Agricultural Economics Report, no. 530. Washington, DC: Economic Research Service, U.S. Department of Agriculture, 1985.

The authors highlight the diversity of the farm sector, concluding that there is growing concern that the farm sector has grown so diverse that a single farm policy may be insufficient to address its needs. "Domestic and international economic policies play important roles in the well-being of farmers, and future farm policy will need to incorporate those concerns if it is to address the issue of instability of incomes and prices."

*** 1.4 ***
Lin, William, George Coffman, and J.B. Penn. *U.S. Farm Numbers, Sizes, and Related Structural Dimensions: Projection to Year 2000.* Technical Bulletin, no. 1625. Washington, DC: U.S. Economics and Statistics Service, U.S. Department of Agriculture, 1980.

In this report, the authors project that the number of U.S. farms will continue to decline through the end of the century--from 2.9 million in 1974 to 1.8 million in 2000. The proportions of small and large farms will change, with large farms increasing and dominating agricultural production. They conclude that "Farm production, farm land, and farm wealth will become more concentrated; farm operators will rent more of their farm land and will produce more of their commodities under contractual arrangements with food processors. The projections are based on four analytical methods: trend extrapolation, negative exponential functions, Markov process, and age, cohort analysis."

*** 1.5 ***

McDonald, Thomas, and George Coffman. *Fewer, Larger U.S. Farms by Year 2000--and Some Consequences*. Agriculture Information Bulletin, no. 439. Washington, DC: Economics and Statistics Service, U.S. Department of Agriculture, 1980.

The authors conclude that "The number of U.S. farms will likely decline by about a third in the next 20 years, if present trends continue, while the number of large farms (annual sales more than $100,000) will quadruple. Family farms will continue to dominate, but the influence of small farms will wane. More large farms will probably mean more farm corporations, more specialization in what farms produce, agricultural production concentrated among relatively few farms, and fewer young people getting started in farming because of high capital requirements."

*** 1.6 ***

Owen, Wyn F., ed. *American Agriculture, the Changing Structure*. Lexington, MA: DC Heath and Company, 1969. LC 74-10655.

The editor presents a collection of 12 essays identifying, from a number of perspectives, important trends and issues relating to the structure of agriculture in the United States. Major sections include (1) the dynamics of agricultural structure, (2) the changing farm firm, (3) the off-farm farmers and agribusiness relations, and (4) perspectives on the future. The materials presented provide the reader with a good representation of the structural characteristics of a highly productive agricultural economy at a given stage of development.

*** 1.7 ***

Reimund, Donn A., Nora L. Brooks, and Paul D. Velde. *The U.S. Farm Sector in the Mid-1980s*. Agricultural Economics Report, no. 548. Washington, DC: Economic Research Service, U.S. Department of Agriculture, 1986.

The authors show that the number of U.S. farms with sales above a quarter million dollars increased by nearly 1 1/2 times over the last decade, but about half the gain was due to inflation. A number of variables, income, wealth, ownership, organization, and concentration of production, are adjusted for inflation to document the actual change between 1974 and 1982. They also present economic profiles of typical farms by region for major commodities.

*** 1.8 ***

Reimund, Donn A., Thomas A. Stucker, and Nora L. Brooks. *Large-Scale Farms in Perspective*. Agriculture Information Bulletin, no. 505. Washington, DC: Economic Research Service, U.S. Department of Agriculture, 1987.

The number of very large farms, farms with annual sales of $500,000 or more, increased from 11,400 in 1974 to 27,800 in 1982. Although these were only 1.2 percent of all farms, they controlled over 10 percent of the land in farms in 1982 and accounted for nearly 33 percent of the total value of farm production. The authors examine recent trends in the number of very large farms, the proportion of total farm land under their control, and their contribution to total output. They also discuss the origin of these farms and their future role in the farm economy.

*** 1.9 ***

Schertz, Lyle P., et al. *Another Revolution in U.S. Farming?* Washington, DC: U.S. Department of Agriculture, 1979.

In this publication, the authors assemble, refine, synthesize, and present available knowledge about how production of livestock and crops is organized and managed, why this is so, how resources are likely to be organized and managed in the future, why, and with what results. The book is based on the concept that the transformation underway in U.S. farming is giving rise to many issues generated by a variety of forces. The materials contained in the book can contribute to an enlightened dialogue about related issues. The premise is that increased public awareness of these changes will lead to more serious consideration and review of current and possible public policies dealing with how resources are organized and managed to produce food and fiber.

*** 1.10 ***

United States. Senate. Committee on Agriculture, Nutrition, and Forestry. *Farm Structure: A Historical Perspective on Changes in the Number and Size of Farms*. 96th Congress, 2d Session. Washington, DC: U.S. Government Printing Office, 1980.

In this set of papers on the continuing issue of farm structure and the effects of farm programs on the control and organization of resources used in farm production, individual authors present a broad perspective, discuss factors of major importance concerning farm structure, and give some of the reasons and implications of changes in farm structure by commodity. The concern of the Congress, which set the process of appraisal in motion, was expressed in the Food and Agriculture Act of 1977. As stated in section 102 (a) of the 1977 Act, "...the maintenance of the family farm system of agriculture is essential to the social well-being of the Nation and the competitive production of adequate supplies of food and fiber."

Diversity and Specialization Among Farms

*** 1.11 ***

United States. Congress. Congressional Budget Office. *Diversity in Crop Farming: Its Meaning for Income-Support Policy.* Special Study. Washington, DC: U.S. Government Printing Office, 1985.

Crop farms vary widely in size and income. In 1982, about 652,000 farms sold about 95 percent of all corn, soybeans, wheat, cotton, rice, and grain sorghum. Among these farms, sales ranged from an average of $70,000 for small farms to $454,000 for those in the largest size category. Farm incomes also varied, from less than $14,000 on the average small farm to more than $95,000 on the largest. Corn and wheat farms had net cash farm incomes that were about average for all crop farms; soybean and sorghum farms had incomes that were below the average; and cotton and rice farms had incomes much above the average.

Farm Financial Problems

*** 1.12 ***

Barry, Peter J. *Financial Stress in Agriculture: Policy and Financial Consequences for Farmers.* AE-4621. Urbana: Department of Agricultural Economics, University of Illinois, 1986.

Barry provides a comprehensive information base to use in evaluating the financial consequences of various policy options and strategies that farmers might follow in responding to the financial stresses of the 1980s. The analysis is based on a projection of changes in financial performance for a set of representative farms over a four-year transitional period. The range of farm types evaluated was broad in terms of geographic coverage (thirteen states) and structural characteristics, farm size, tenure position, enterprise mix, and asset composition. The mixture of farm types included cash grain farms, livestock farms, two dairy farms, and one ranch.

*** 1.13 ***

Barry, Peter J., Paul N. Ellinger, and Vernon R. Eidman. "Firm Level Adjustments to Financial Stress." *Agricultural Finance Review* 47 (Special Issue, 1987):72-99.

The authors report the results of a simulated financial analysis of alternative farm types in a series of charts "as an information base for use in evaluating the financial consequences of various policy options and strategies that farmers and lenders might follow in responding to financial stress."

*** 1.14 ***

United States. Comptroller General. *Financial Condition of American Agriculture.* Report to the Congress, GAO/RCED-86-09. Washington, DC: U.S. Government Printing Office, 1985.

In this comprehensive report of the effects of farm debt expansion from 1970 to 1981 on prices of farm land and some of the related effects on farm incomes, the comptroller general observed that "As gross income leveled off and real interest rates rose in the 1980s, farmers' abilities to meet their debt payments began to deteriorate. Borrowing against declining asset values became more difficult, and farm sector stress intensified. The Department of Agriculture has estimated that of the 636,000 commercial-size farms, 31,000 are insolvent; that is, their debts exceed assets. A total of 197,000 farms are financially stressed; that is, their debt/asset ratios exceed a level (40 percent) where most farms can be profitable. There is some geographical concentration, with 60 percent of financially-stressed farms located in just 12 midwestern states."

*** 1.15 ***

Hall, Bruce F., and E. Phillip LeVeen. "Farm Size and Economic Efficiency: The Case of California." *American Journal of Agricultural Economics* 60:4 (November 1978):589-600.

Hall and LeVeen examine the relationship between farm size and production costs and conclude "that relatively modest-sized farms can achieve a major portion of the possible cost savings associated with size." They analyze the sources of efficiency and show that factors other than labor-saving technology may be important contributors to economic efficiency. The implications of the analysis were developed for the debate over acreage restrictions in reclamation policy. Hall and LeVeen concluded that "Strict enforcement of the 160-acre limit could cause a modest overall efficiency loss, but this would be borne by landowners rather than consumers."

*** 1.16 ***

Hansen, Gregory D., and Jerry L. Thompson. "A Simulation Study of Maximum Feasible Farm Debt Burdens by Farm Type," *American Journal of Agricultural Economics* 62:4 (November 1980):727-733.

Hansen and Thompson conclude that "Financial leverage, which magnifies the effect of farm income instability upon 'bottom line' productivity, becomes less feasible as income fluctuation increases... With a flexible re-payment agreement and with land valued at current prices, substantial debt use was feasible for many farm types in the years 1966-75.... Management ability appeared to have a greater influence upon debt capacity than farm size. Debt-servicing ability was improved more by becoming a good manager than simply by becoming a large operator."

*** 1.17 ***

Harrington, David H., Donn A. Reimund, Kenneth H. Baum, and R. Neal Peterson. *U.S. Farming in the Early 1980's: Production and Financial Structure*. Agricultural Economic Report, no. 504. Washington, DC: Economic Research Service, U.S. Department of Agriculture, 1983.

The authors update information on organizational trends in farming and some of the forces shaping those trends, and assess the outlook for selected types of farms under alternative economic situations. They conclude that "Long-term trends to fewer and larger farms appear to have slowed sharply in the late seventies and early eighties. High leverage of farm assets, declining asset values, and low commodity prices caused severe cash flow problems for the farm sector in the early eighties, and many farmers incurred sizable losses. Computer simulation models of representative farms in six regions of the country suggest that fully owned farms with modest or no debt should fare reasonably well over the next few years. But part-owner farms and heavily indebted farmers will likely face financial difficulty and declining net worths. Commodity programs were found to contribute significantly to the survival ability of farms."

*** 1.18 ***

Johnson, James D., Mitchell J. Morehart, and Kenneth Erickson. "Financial Conditions of the Farm Sector and Farm Operators." *Agricultural Finance Review* 47 (Special Issue, 1987):1-20.

In this paper, the authors present estimates of the farm sector's income and balance sheet to provide a perspective on the financial performance of farm businesses, returns earned by resources used in farm production, and earnings of individuals and households who either invest in farms or are employed in farming. Aggregate sector estimates also provide a foundation for understanding the diversity that exists in the financial circumstances of farmers.

*** 1.19 ***

Lewis, James A. "Landownership in the United States, 1979." *Agricultural Information Bulletin*, no. 435. Washington, DC: Economics, Statistics, and Cooperatives Service, U.S. Department of Agriculture, 1980.

Lewis' report is based on a 1978 survey of landowners in the United States, conducted by the Natural Resource Economics Division (NRED) of the Economics, Statistics, and Cooperatives Service (ESCS), U.S. Department of Agriculture. The survey data provide a set of owner characteristics, such as age, income, occupation, race, and education, for the survey year, and basic information on the quality and use of land, which can be linked to the past decision patterns of owners who made certain types and levels of investments or who made no change in land use.

*** 1.20 ***

Lowenberg-DeBoer, J., and Michael Boehlje. "The Impact of Farmland Price Changes on Farm Size and Financial Structure." *American Journal of Agricultural Economics* 68:4 (November 1986):838-848.

A modified Vickers model was used to show that farm land capital gains provide incentive to increase farm acreage and debt use. Farm land capital losses have the opposite effect. The model indicates that part of the current financial vulnerability of farms can be traced to management decisions made in response to the farm land capital gains of the 1970s. The effects are not purely tax driven, though taxes can affect the magnitude of incentives. The Vickers model was modified to allow a finite horizon, taxes, and the recognition of unrealized capital gain or loss.

*** 1.21 ***

Prentice, Paul T., and David A. Torgerson. "U.S. Agriculture and the Macroeconomy." *Agricultural-Food Policy Review: Commodity Program Perspectives*. Agricultural Economic Report, no. 530. Washington, DC: Economic Research Service, U.S. Department of Agriculture, 1985.

The authors show that the farm economy and the general economy are so closely linked that economic conditions and policies beyond the farm gate can affect farmers' well-being as strongly as farm programs for individual commodities. Macroeconomic conditions and policies affect demand for farm products and thus farmers' revenue as well as the cost of farming. They conclude that "Long run trends in the general economy suggest that future growth in domestic demand will not be sufficient to eliminate excess farm production. A macroeconomic policy mix of fiscal stimulus combined with monetary restraint is harmful to agriculture and other interest-sensitive, export-dependent, or import-competing sectors--at least in the short run."

*** 1.22 ***

United States. Department of Agriculture. *A Time to Choose: Summary Report on the Structure of Agriculture*. Washington, DC: U.S. Department of Agriculture, 1981.

The summary report is based on the Structure of Agriculture Project, initiated in March 1979, by Secretary Bob Bergland, to research current structural issues to determine the impacts of current market forces and policy on agriculture, and to recommend policy alternatives. It concluded that "unless present policies and programs are changed so that they counter, instead of reinforce and accelerate the trends towards ever-larger farming operations, the result will be a few large farms controlling food production in only a few years."

*** 1.23 ***
United States. Economics, Statistics, and Cooperatives Service. *Structure Issues of American Agriculture.* Agricultural Economic Report, no. 438. Washington, DC: U.S. Department of Agriculture, 1979.

The report, which involved more than 50 members of the staff of the Economics, Statistics, and Cooperatives Service, was designed to assist in considering the structure issues initiated by Secretary of Agriculture Bob Bergland. The purpose of the articles was to provide information that would increase understanding and facilitate discussion on the structure of agriculture. Although most of the discussion is on the economic aspects of structure, the volume also treats the related demographic, sociological, and historical aspects. Thus it represents the first coordinated attempt by social scientists in the Department of Agriculture to address this issue from the perspective of several disciplines.

*** 1.24 ***
United States. Economic Research Service. *Financial Characteristics of U.S. Farms, January 1985.* Agricultural Information Bulletin, no. 500. Washington, DC: U.S. Department of Agriculture, 1986.

The USDA offers a broad and comprehensive report, giving fairly complete and detailed data on the subject of financial characteristics. Hence, it serves as a source of data for other informative or analytical studies.

*** 1.25 ***
United States. Congress. General Accounting Office. *Changing Character and Structure of American Agriculture: An Overview.* CED-78-178. Washington, DC: U.S. Government Printing Office, 1978.

In this report, the GAO concludes that the farm sector is heading toward fewer but larger operating units, which are capturing a larger portion of total farm sales. In an attempt to maintain income through increased production, farmers have made use of technological breakthroughs. However, they have found themselves requiring more equipment and then more land, and still more powerful and faster equipment to stay ahead of narrowing profit margins, inflation, and competitive pressures. More specifically the report also concludes that "The result of farm product specialization over the last two decades was that farm worker productivity increased nearly twice as fast as that of the industrial worker. However, to maintain this productivity, the farmer became dependent upon petroleum-based inputs of fuel, fertilizer, and pesticides as well as other agro-industrial services to operate his increasingly specialized farm. As these specialized and nonrenewable inputs become more valuable, cost/price inflationary pressures on the farmer will increase."

General Works

*** 1.26 ***
Cramer, Gail L., and Clarence W. Jenzen. *Agricultural Economics and Agribusiness: An Introduction.* New York: John Wiley and Sons, 1979. ISBN 0-471-04429-6.

Cramer and Jenzen's basic text is designed for a one-quarter or one-semester class in an introductory agricultural economics or agribusiness course. The book explores the structure and organization of agriculture and discusses economic principles as they apply to agriculture.

*** 1.27 ***
Halcrow, Harold G. *Economics of Agriculture.* New York: McGraw-Hill Book Company, 1980. ISBN 0-07-025556-3.

Halcrow's textbook is designed for a foundation course applying principles of economics to agricultural production, marketing, and public policy. It begins with the basic concepts of economics and agriculture, develops a broad view of the central economic problems of agriculture, and applies the concepts and principles to show how these problems are solved.

*** 1.28 ***
Hamilton, Carl. *In No Time at All.* Ames: Iowa State University Press, 1974. ISBN 0-8138-0825-1.

Hamilton recalls the familiar trappings of country life in the midwest between the world wars, and remembers and shares hundreds of recollections, to create a humorous and honest piece of nostalgia. The images of family farms from 1910 to 1940 contrast sharply with those of the current era.

*** 1.29 ***
Hayenga, Marvin, V. James Rhodes, Jon A. Brandt, and Ronald E. Deiter. *The U.S. Pork Sector: Changing Structure and Organization.* Ames: Iowa State University Press, 1985. ISBN 0-8138-1396-4.

The authors offer one of a series of commodity sector studies stemming from the North Central Region Research Project NC-117 (Studies of the Organization and Control of the U.S. Food System). It describes the organizational structure of each vertical stage of the pork sector, describes the current pricing and coordination systems prevalent at each stage of the vertical market system, and evaluates performance of the pork sector and any public policy options that might improve sector performance.

*** 1.30 ***

Heady, Earl O. *Economics and Social Conditions Relating to Agriculture and Its Structure to Year 2000.* Ames: Iowa State University Press, 1980.

Heady makes use of a wide range of knowledge and computer technology to project conditions to the year 2000. The methodology is of special importance for students and researchers. Many of the conclusions will be of interest to students as well as more general readers.

*** 1.31 ***

Kramer, Mark. *Three Farms Making Milk, Meat, and Money from the American Soil.* New York: Bantam Books, 1981. ISBN 0-316-50315-0.

Kramer's study is a series of portraits of some of the human consequences of the shift from the traditional family farm to modern large-scale, high technology farming systems. It is a sad, funny, and incisive book, which blends perceptive reporting with lively and provocative writing.

*** 1.32 ***

McMillen, Wheeler. *Feeding Multitudes: A History of How Farmers Made America Rich.* Danville, IL: Interstate Printers and Publishers, n.d. ISBN 0-813421-926.

McMillen's objective in this book is "to present a fairly comprehensive, non-academic narrative of how productive farmers enriched the United States from 1607 to the bicentennial era." Because of its narrative format, the book is a useful information source for students of U.S. history.

*** 1.33 ***

Schultz, Theodore W. *The Economic Organization of Agriculture.* New York: McGraw-Hill Book Company, 1952. LC 52-8323.

Schultz has written one of his largest and most comprehensive books. It draws on a large amount of data dealing with organization and structure to illustrate the dynamic nature of agriculture and the changes that occur over time.

*** 1.34 ***

Schwenke, Karl. *Successful Small-Scale Farming.* Charlotte, VT: Garden Way Publishers, 1979. ISBN 0-88266-148-5.

Schwenke argues that small-scale farming can be successful, gives a number of suggestions for such farming, and provides examples of successful small-scale farms. He discusses some suggested uses of technology, capital, and labor on small farms.

CHAPTER 2

THE SCIENTIFIC AND TECHNOLOGICAL REVOLUTION IN FARMING

"The capacity to develop and manage technology in a manner consistent with a nation's physical and cultural endowments is the single most important variable accounting for differences in agricultural productivity among nations."

Vernon W. Ruttan (1982), *Agricultural Research Policy.*[1]

Technology is the key to productivity of the American farm and our competitive position in the world. A myriad of new technological advances will dominate growth prospects for the American farm over the next generation. How these technologies are developed, used, or managed is a crucial issue.

Technology is the key to productivity of the American farm, its growth and size distribution, and its ability to compete in the global economy. Since 1940, significant increases in productivity have come from successfully applying new technologies that are cost reducing and output increasing. A myriad of new technological advances will dominate growth prospects for the American farm at least over the next generation.

How these technologies are to be developed, and used or managed, is a crucial issue for the American farm and the population that depends on it. Advances in technology that increase productivity generally require major adjustments in the structure of the farm economy. Gains to successful farm innovators, consumers, and national welfare must be evaluated as we consider the adjustments required on the American farm.

Technology and Productivity

The term technology generally implies "a specific state of art and science which is used to transform a set of inputs (resources) into a set of outputs (goods or services)." Thus the American farm uses a large number of technologies in production. Many of these are supplied by farm input industries producing machinery, chemicals, and other inputs used on the farm. Another set of technologies is used in processing and marketing farm products. New technology also involves advances in organization and management at all levels among input industries, farms, and processing-marketing firms.

Technological change occurs when either the

Table 2-1.

Indexes of the American Farm's Total Input, Total Output, and Productivity (1977 = 100).

Year	Total Input	Total Output	Productivity
1940	97	50	52
1950	101	62	61
1960	98	75	77
1970	97	84	87
1980	103	104	101
1985	100	127	127

Source: Economic Research Service, United States Department of Agriculture, *Outlook 1987 Charts.* (Washington, DC, 63rd Annual Agricultural Outlook Conference, December 1986), 102, 103; and *Agricultural Outlook, 1987 Outlook Issue.* (Washington, DC, January-February 1987), 49.

form or mix of inputs, or the form or mix of outputs, are modified from their previous state. Some changes are very important, such as the change from horses to farm tractors, the introduction of hybrid corn, and the modern improvements in farm chemicals. The increase in productivity on the American farm does not depend on just a few of these, but on a myriad of scientific and technological advances.

The term productivity, when used appropriately, refers to the ratio of output to input. Thus an advance in productivity means an increase in output *relative* to input, and such an advance may refer to either *total* or *partial* productivity. Total productivity refers to output per unit of total input. In measuring the productivity of the American farm, the inputs used in production--land, labor, buildings and equipment, energy, fertilizers, chemical pesticides, and others--are value weighted to yield an index number of total input, or inputs. An index of output is constructed in similar fashion by value weighting all farm products. Using a given year as a base, an index of productivity can be constructed for any year by dividing the index of total output by the index of total input (Table 2-1). Also, the average annual rates of change can be computed for total input, total output, and productivity (Table 2-2).

The total productivity measure cannot capture all factors that are important, such as the effect of technological change on the environment (off-farm pollution for example), or social costs such as the forced liquidation of family farms. So technology evaluation cannot be limited just to productivity measures, but also must take other things into account, such as the environment and some broader economic and social effects, both on and off the farm.

Partial productivity measures the ratio of output to a selected input, or group of inputs. Some frequently used ratios refer to land, to animals, or to farm labor. In the past half century, for instance, corn yields in the United States have increased from less than 40 bushels to well over 100 bushels per acre. Since 1950, the total number of dairy cows has been reduced by one-half and dairy cows are producing more milk on one-third less feed. Since 1800, the hours of farm labor on the American farm required to produce 100 bushels of wheat have dropped from 373 hours to less than nine hours. Partial productivity relates to any input or group of inputs, such as tractors or farm power, a class of farm machinery or all farm machinery.

Over the last half century (as shown in Tables 2-1 and 2-2), increasing productivity, rather than use of more inputs, has been the source of growth in farm output. From 1900 to 1925, productivity declined, while growth in output depended entirely on using more inputs. Over a longer period, from 1880 to 1940, average grain yields changed scarcely at all.[2] Increases in livestock output depended mainly on using more land for grazing, growing more feed crops, and raising more animals. Beginning in the 1930s, with the spread of hybrid corn and

Table 2-2.

Average Annual Rates of Change (percent-per-year) in the American Farm's Total Input, Total Output, and Productivity, 1900 to 1985.

Item	1900-25	1925-50	1950-65	1965-79	1980-85
Total input	1.1	0.2	-0.4	0.3	-0.6
Total output	0.9	1.6	1.7	2.1	4.6
Productivity	-0.2	1.3	2.2	1.8	5.2

Source: United States Department of Agriculture, Economic Research Service, *Agricultural Outlook* and *Agricultural Outlook Charts* (various years).

cheaper fertilizer, total farm productivity began to increase steadily. This trend continued through the next four decades,[3] until in the 1980s the American farm entered a new age of high technology.

The New Age of High Technology[4]

The American farm is on the threshold of a new age of high technology, involving the sciences of biotechnology and information technology. Biotechnology includes any technique that uses living organisms to make or modify products, to improve plants or animals, or to develop microorganisms for specific uses. New information technologies involve significant improvements in communication and information management for organizing and monitoring farm production processes, marketing, and other activities. Developing together, biotechnology and information technology will accelerate growth in productivity of the American farm far beyond what has been experienced before.

Efficient application of biotechnology on the American farm promises significantly more rapid increases in productivity in the next generation than has occurred previously. The most dramatic effects will be felt first in dairying, where new genetically engineered phamaceuticals (such as bovine growth hormones and feed additives) and information management systems (using computers and automatic controls over feed inputs) are being introduced commercially. Milk production per cow, for instance, under the most likely environment, will increase 3.9 percent annually over the next several years, as compared with 2.6 percent annually from 1965 to 1985. Productivity of most other farm animals will also increase through advances in reproduction and feed technologies.

Changes induced by genetic engineering in animal reproduction will involve three general areas, or methods: (1) recombinant (rDNA) techniques, also called "gene splicing," (2) monoclonal antibody production, and (3) embryo transfer.

Because of the power of rDNA technology to alter life forms of plants and animals (including humans), it is regarded as one of the greatest--if not the greatest, and also most controversial--achievements of biological science. Basic to this technology, molecules of rDNA formed from two different species, can be inserted into a variety of bacteria, yeasts, and animal cells, where they replicate and produce many useful proteins and other nutrients like amino acids and single-cell protein feed supplements.

One of the applications of these new pharmaceuticals is to manufacture growth hormones that can be injected into animals to increase their productive efficiency. Injections of bovine growth hormone (bGH) into dairy cows, for instance, at the rate of 44 milligrams per cow per day, have resulted in increases of 10 to 40 percent in milk yield. Of greater importance for new generations, genes for new traits to be inserted into the reproductive cells of livestock and poultry, thus permanently endowing future generations with the desired traits, such as disease resistance or growth, having far-reaching consequences.

Monoclonal antibody production can create a large volume of a single antibody, which can be produced by fusing a myeloma cell (a cancerous antibody producing cell) with a cell that yields an antibody. This creates a hybridoma, which in turn produces (theoretically and in perpetuity) large quantities of identical (i.e. monoclonal) antibodies. The many important uses of monoclonal antibodies that will increase farm productivity, as well as having other much broader consequences, include purifying proteins made by rDNA; passive immunization of calves against scours; substitutions for vaccines, antitoxins, and antivenoms; sexing of livestock embryos; post-coital contraception and pregnancy testing; detecting food poisoning; imaging, targeting, and killing of cancer cells; monitoring of levels of hormones and drugs; and preventing rejection of organ transplants.

Embryo transfer is used for rapidly upgrading the quality and productivity of livestock, particularly cattle. In the process, a super-ovulated donor animal is artificially inseminated, and the resulting embryos are removed nonsurgically for implantation in, and carrying to term by, surrogate mothers. Before implantation, the embryos can be sexed with a monoclonal antibody, split to make twins, fused with embryos of other animal species, or frozen in liquid nitrogen for storage until the estrous cycle of the surrogate mother is in synchrony with that of the donor. Vigorous pursuit of research in these areas could result, by year 2000, in marketing large numbers of genetically improved embryos containing genes that will improve fertility and fecundity. This will result in improved rates of gain among meat animals with lower body fat, improved carcass characteristics, increased milk production, and increased resistance to disease among offspring.

Other new products will be used to diagnose and treat diseases that have reduced the productivity of livestock and poultry on the American farm by an estimated 20 percent annually. For instance, a genetically engineered vaccine for colibacillosis (scours), a disease that has killed millions of newly-born calves and piglets each year, has recently become available commercially. A subunit vaccine has been genetically engineered for foot-and-mouth disease in cattle, which is still a serious disease throughout Africa, South America, and parts of Asia. Viral infections will be reduced by advances in using interferons (certain glycoproteins--proteins bound to sugar groups) that regulate the body's immune response. Genetic engineering of the micro-organisms that permit ruminants--cattle, sheep, and goats--to digest forages may permit these livestock to digest various feedstuffs, such as high-lignin substances, that up to now have been virtually useless to them.

Biotechnology for plants is developing more slowly than for animals; the most significant impacts on plant productivity will not be felt before the turn of the century. After 2000, the impacts on plants will be substantially greater than on animals. Applications of biotechnology can modify crops so that they will make more nutritious protein, resist insects and disease, grow in harsh environments, and make their own nitrogen fertilizer. Genetic engineering applications will include microbial inocula, in vitro plant propagation; and genetic modification.

Plant propagation techniques, which have been used extensively to produce new species and varieties, such as hybrids for example, are being extended to new areas. The efficiency of photosynthesis, the fundamental basis for plant growth, will be enhanced over several years as a result of research on several factors that naturally limit the efficiency of the process.

New information technologies in various stages of rapid development will greatly improve management and increase productivity of the biotechnologies. On-farm communication and information technologies may include a central computer system, a radio link with operators of tractors, combines, and other implements, and an on-farm weather station. The central computer system may include remote terminals with keyboards, display screens, and printers for entering data to be used by the farm operator. Many processes in plant and animal production will be organized and monitored by new electronic technologies. Some are designed to provide more information to the farm operator. Others will operate automatically, based on a continuing flow of information, such as devices to regulate the flow of irrigation water, monitor and control pests, and regulate the environment for livestock. Other applications include devices to identify individual animal performance for feeding, reproduction, and genetic improvement. New and improved computer technology will produce significant changes in insect and mite management at all levels. New systems will provide rapid analysis of pesticide and field management information, reports of new or unknown pests, general pest survey information, and specified field locations with pest severities.

Telecommunication systems will provide information that is crucial for management decisions, and also link the farm with other firms and institutions with which the farm operator communicates. These will include both voice and data systems, enabling farmers to have rapid, inexpensive, and reliable access to a wide array of information, such as markets, weather, and other subjects of interest. Satellite-based communication technologies will provide more accurate and timely information. Remote sensing technology will be used increasingly to detect, process, and analyze a wide range of data collected from a distance. Weather forecasting, based on remote sensing technology, has already made a great impact on farm production by helping farmers to decide when to plant and harvest; weather modification is possible in the future. Remote sensing technologies will provide timely information on crop prospects over wide areas, and give an array of information useful for managing and marketing decisions.

Human Capital, Welfare, and Environment

Taking the long view, the critical component in the age of high technology and modernization is human capital: the productivity of people and their welfare, or environment. It is critical to recognize the value of education and other investments in people, and the adjustments that people must make as they adapt to the new age. Although the farm population has dropped by more than two-thirds from its high point in the 1930s, it could drop by another 50 to 75 percent while still increasing total farm productivity and output. The farm population needs more help in planning what to do and in carrying through on plans that are made.

The new technology will enhance the productivity of people on farms, protect the natural resources such as land by diminishing the relative importance of farm land,[5] and enhance the welfare of consumers by bringing a more abundant and still cheaper food supply. Such advances cannot be made, however, without dealing with the many adjustments that must be made by farm families and farm-related industries. Also, there are other concerns directly affecting consumers. For instance, there are concerns about the effects on human health of the hormones used to increase growth rates of livestock and the possible contamination of food by pesticides used in producing field crops, especially some fruits and vegetables. Although very little has been proven, there are concerns that some of the growth hormones used on livestock might contribute to cancer in humans; this is a matter of continuing research. Concerns about pesticide residues in foods apply to crops produced in the United States, as well as to food that is imported, especially fruits and vegetables imported from Mexico and other countries in Central America.

In the United States, pesticide manufacture, sale, and use is regulated and controlled by both federal and state laws and regulations. The Federal Insecticide Act of 1910 was the first general law passed to control the manufacture, sale, and transportation of insecticides and fungicides. Additional acts, especially in 1947, 1954, and 1972, greatly strengthened regulations concerning manufacture, sale, and use of pesticides. The federal Environmental Pesticide Control Act of 1972, which replaced all previous acts, required all pesticides to be classified for general or restricted use, and made unlawful the sale or use of any pesticide inconsistent with its labeling. Some pesticides cannot be manufactured or sold, and those in a restricted category can be used only by or under the supervision of certified applicators, or subject to such other restrictions as the administrator of the Environmental Protection Agency (EPA) may determine. As new and more effective pesticides are manufactured and sold, more stringent environmental regulations and stronger enforcement measures will be required. What should be done is properly the subject of ongoing research at both federal and state levels.[6]

A more difficult problem of control is involved in regulating imported fruits and vegetables. Pesticide use has been expanding every year in the exporting countries. Inspection of growing conditions is not under control of U.S. government officials, and detecting possible contamination in the volume of imports is difficult, slow, and costly. Corrective measures are often too slow to protect U.S. consumers.[7] But, since there is a strong demand in the United States during the winter months for imported fruits and vegetables, new ways must be found to assure that these imports are free from pesticide residue. Given success in this effort, and with continuing improvements in U.S. production, a safe and nutritious food supply can be assured.

The new age of high technology is bringing new challenges for management, which is the subject of the next chapter. After that we turn to study marketing, the effects of farm adjustment on rural communities, and alternatives in farm policy.

Endnotes

1. Vernon W. Ruttan. *Agricultural Research Policy*. Minneapolis: University of Minnesota Press, 1982. p.17.

2. D. Gale Johnson, and Robert L. Gustafson. *Grain Yields and the American Food Supply, an Analysis of Yield Changes and Possibilities*. Chicago: The University of Chicago Press, 1962. p.8.

3. For a more extensive discussion and additional sources of information, see Harold G. Halcrow. *Food Policy for America*. New York: McGraw-Hill Book Company, 1977. pp. 91-133.

4. United States. Congress. Office of Technology Assessment. *Technology, Public Policy and the Changing Structure of American Agriculture.* OTA-F-285. Washington, DC: Government Printing Office, 1986. This publication, which contains numerous references to scientific papers and other reports, is the major source of material for the section on the new age of high technology. This section also benefits from an unpublished paper by John R. Campbell, dean of the University of Illinois College of Agriculture, "Views on the Future of Agriculture," January 31, 1986.

5. Harold G. Halcrow, Earl O. Heady, and Melvin L. Cotner, eds. *Soil Conservation Policies, Institutions and Incentives.* Ankeny, IA: Published for North Central Research Committee 111: Natural Resource Use and Environmental Policy by the Soil Conservation Society of America, 1982. pp. 201-282.

6. Harold G. Halcrow. *Food Policy for America.* New York: McGraw-Hill Book Company, 1977. pp. 109-133.

7. *The Wall Street Journal*, March 26, 1987.

Chapter 2: Annotated Bibliography

Technology and Productivity

*** 2.1 ***

Antle, John M. "The Structure of U.S. Agricultural Technology, 1910-78." *American Journal of Agricultural Economics* 66:4 (November 1984):414-421.

Antle uses 1910-78 time-series data and a single product aggregate translog profit function to measure the structure of U.S. agricultural technology. Duality relations are used to devise a multi-factor measure of biased technical change. A measure of non-homotheticity is introduced that indicates the effects of changes in scale on aggregate cost shares. He concluded that "although different, non-homothetic technologies characterized the pre-war and post-war periods, differing technical changes are consistent with relative price trends during the two periods."

*** 2.2 ***

Capalbo, Susan M., and Michael G.S. Denny. "Testing Long-Run Productivity Models for the Canadian and U.S. Agricultural Sectors." *American Journal of Agricultural Economics* 68:3 (August 1986):615-625.

The authors' purpose in this paper was to establish linkages between the gross and net productivity indexes and the implied production structures. These linkages were developed as separability restrictions on various subgroups of inputs and time in a general production model. The restricted models were arranged in sequences for nested hypotheses testing. The authors concluded that "Empirical evidence from the U.S. and Canadian agricultural sectors based on a translog production function supports the gross output total factor productivity structure. The net output Hicks neutrality hypothesis is rejected for both the United States and Canada."

*** 2.3 ***

Eckstein, Zvi. "The Dynamics of Agricultural Supply." *American Journal of Agricultural Economics* 67:2 (May 1985):204-214.

Eckstein studied the linear rational expectation models that have been developed recently in literature dealing with macroeconomics. He concluded that they proved "to provide a useful methodology for analyzing and estimating agricultural supply elasticities as well as for interpreting the cyclical behavior of agriculture markets."

*** 2.4 ***

Offutt, Susan E., Philip Garcia, and Musa Pinar. "Technological Advance, Weather, and Crop Yield Behavior." *North Central Journal of Agricultural Economics* 9:1 (January 1987):49-64.

The authors explored the relationships between yield level stability and advances in technology and changes in weather, with application to U.S. corn production. Based on evaluation at the farm, sub-state, and national levels, they found no evidence of yield plateaus. Absolute, but not relative, yield variability net of technology time trend alone was seen to have increased over time. The authors concluded that "When yield behavior is adjusted for weather, variances are more likely to be equal between the two periods. These results suggest that technology is not the only determinant of changing yield risk."

*** 2.5 ***

Smith, Edward G., Ronald D. Knutson, and James W. Richardson. "Input and Marketing Economies: Impact on Structural Change in Cotton Farming on the Texas High Plains." *American Journal of Agricultural Economics* 64:3 (August 1986):716-720.

The authors isolated the degree of pecuniary

economies existing for input purchases and marketing on cotton farms in the Texas Southern High Plains. They concluded that "substantial cost and marketing economies are being realized by the largest farms in the region."

*** 2.6 ***

Stucker, Thomas A., Richard F. Fallert, and Kathryn L. Lipton. "Bovine Growth Hormone Brings Progress to Dairy Farms." *National Food Review, the Food Industry: Changing with the Times.* NFR-35. Washington, DC: Economic Research Service, U.S. Department of Agriculture, 1986.

The authors observed that "the union of biology and technology is propelling agriculture into a new era of advancement," and they concluded that "An implication of bGH, bovine growth hormone, like the technologies that came before, is the prospect of lower prices and perhaps fewer farmers. However, the potential benefits of bGH are greater efficiency, lower costs of production, the essential ability to compete in the world dairy market and with substitute dairy products, and increased consumption at lower consumer prices."

*** 2.7 ***

Teigen, Loyd D., Felix Spinelli, David H. Harrington, Robert Barry, Richard Farnsworth, and Clark Edwards. *The Implications of Emerging Technologies for Farm Programs.* Report no. 530, 54-72. Washington, DC: Economic Research Service, U.S. Department of Agriculture, 1985.

The authors show that technological innovation is important in the growth and development of the farm economy. "Resources freed through adoption of new technologies can be put to other productive uses. Technology that promotes efficiencies in U.S. agriculture is important to consumers and to the competitive position of U.S. farmers in international markets. However, the adoption of new technologies can have structural and distributional implications for the farm sector."

The New Age of High Technology

*** 2.8 ***

Burt, Oscar R. "Econometric Modeling of the Capitalization Formula for Farmland Prices." *American Journal of Agricultural Economics* 68:1 (February 1986):10-26.

Burt formulated an econometric model to explain the dynamic behavior in farm land prices. A second-order rational distributed lag on net crop-share rents received by landlords captured the dynamic movements of prices and performed well in conditional post-sample forecasts. He concluded that "The adjustment path of land prices in response to a perturbation in rents is a protracted dampened cycle. The implicitly estimated tax-free capitalization rate on rent associated with equilibrium land price is 4.0%. Neither the expected rate of inflation nor an exponential trend on rent expectations has a significant effect on land prices."

*** 2.9 ***

Butler, L.J., and A. Allan Schmid. "Genetic Engineering in the Future of the Farm and Food System." *The Farm and Food System in Transition*, no. 15. East Lansing: Michigan State University. Cooperative Extension Service, 1984.

Butler and Schmid discuss some of the potential changes in agriculture, agribusiness, and the universities resulting from genetic engineering. They place emphasis on institutional structures and policy issues, which cause technological change, as well as on those that react to such change.

*** 2.10 ***

Decker, Wayne L. "The Implications of Climate Change for 21st Century Agriculture." *The Farm and Food System in Transition*, no. 47. East Lansing: Michigan State University. Cooperative Extension Service, 1983.

Decker notes that "Climate is dynamic, fluctuating from year to year and changing over time periods ranging from geologic eons to centuries to generations to decades." He concludes that "With or without human impact (such changes made in CO_2 concentrations) the agricultural system must be ready to respond to these changes in climate....The institutions for research and extension education are in place, but maintenance of the system will be necessary. Because of the increased stress on plant and animal agriculture, an improved weather information system for agriculture will be needed."

*** 2.11 ***

Johnson, Glenn L., and Sylvan H. Wittwer. *Agricultural Technology until 2030: Prospects, Priorities, and Policies.* Special report, no. 12. East Lansing: Michigan State University. Agricultural Experiment Station, 1984.

Johnson and Wittwer take a broad and comprehensive view of projections for American agriculture. They indicate that international competition in commodity markets, possible energy shortages and foreign exchange needs to buy energy, demands for improved world and U.S. diets, and population growth will make it advantageous for the United States to develop capacity to double agricultural production in the next half-century or so. The targets that they accept as reasonable, feasible, and in the national interest, include creating capacity to: increase yields an average of 40 to 50 percent by 2010, and 65 to 80 percent by 2030. They estimate that this will use an additional 35 to 40 million acres by 2010, and 50 to 60 million more fragile soils in stressed environments for crops and improved forages by 2030. This will use the same or less fossil energy while increasing the use of skilled labor, capital, and expendable inputs to virtually eliminate unskilled farm "stoop" labor. They conclude that, "In addition to better technologies, we will need improved institutions and more skilled people. We would also have to ensure an agricultural social and economic structure that adds to rather than detracts from the quality of life of both rural and urban people."

*** 2.12 ***

United States. Congress. Office of Technology Assessment. *Technology, Public Policy, and the Changing Structure of American Agriculture.* Washington, DC: U.S. Government Printing Office, 1986. LC 85-600632.

In this most comprehensive and detailed report, the OTA addresses "the longer run issues that technology and certain other factors will have on American agriculture during the remainder of this century." It analyzes the relationships of new biotechnology and information technologies on "agricultural production, structural change, rural communities, environment and natural resource base, finance and credit, research and extension, and public policy." The report identifies many of the benefits that new technologies will create, but warns that these benefits will also exact substantial costs in potential adjustment problems. The authors conclude that "with a continued flow of new technologies into the agricultural production system, major crop yields will continue to grow and U.S. agriculture will continue to provide enough food to meet domestic and foreign demand as long as agricultural research is adequately supported and economic and political environments are favorable."

Human Capital, Welfare, and Environment

*** 2.13 ***

Adams, R.M., S.A. Hamilton, and B.A. McCarl. "The Benefits of Pollution Control: The Case of Ozone and U.S. Agriculture." *American Journal of Agricultural Economics* 68:4 (November 1986):886-893.

The authors report on an assessment of the benefits to agriculture arising from reductions in ambient ozone pollution. Estimates are derived using recent plant science data as input for a spatial equilibrium model of U.S. agriculture. Sensitivity of benefit estimates to biological and economic sources of uncertainty is also investigated. The authors' results suggest that "the benefits of a 25 percent reduction in ambient ozone are substantial, amounting to $1.7 billion. The robustness of these estimates varies across alternative assumptions concerning response data and export markets."

*** 2.14 ***

Anderson, Lee G., and Dwight R. Lee. "Optimal Governing Instrument, Operation Level, and Enforcement in Natural Resource Regulation: The Case of the Fishery." *American Journal of Agricultural Economics* 68:3 (August 1986):678-690.

Anderson and Lee noted that most regulation studies have used industry output or inputs as the control variable(s), but that these are only indirectly controlled by government action through its choice of governing instrument, enforcement procedure, and penalty structure and the operational level of each. They developed a model that demonstrates how profit-maximizing firms will react to these control variables, taking into account the benefits (extra production) and costs (possible penalties) of noncompliance and the ability to avoid detection of noncompliance. The optimal operation level for two sets of control variables is derived and discussed.

*** 2.15 ***

Centner, Terence J., and Michael E. Wetzstein. "Reducing Moral Hazard Associated with Implied Warranties of Animal Health." *American Journal of Agricultural Economics* 69:1 (February 1987): 143-150.

The authors noted that implied warranties convey information regarding animal health to livestock buyers, and that statutory implied warranties also include insurance coverage for a warranty breach, which removes buyers' incentive to mitigate possible damages when a warranty is breached. Non-mitigation of damages has created the problem of moral hazard, which has led livestock sellers to seek legislative exemptions abrogating implied warranties. Centner and Wetzstein developed a theoretical analysis of the possible inefficiencies associated with legislative exemptions adopted by twenty-five states, and they suggested partial insurance as an alternative to the legislative shift in liability.

*** 2.16 ***

Karp, Larry, Arye Sadeh, and Wade L. Griffin. "Cycles in Agricultural Production: The Case of Aquaculture." *American Journal of Agricultural Economics* 68:3 (August 1986):553-561.

The authors considered the problem of determining optimal harvest and restocking time and levels. A continuous time deterministic control problem was used to study the case where production occurs in a controlled environment. A stochastic control problem was then used to determine rules for the cultivation of *P. stylirostris,* which occurs in a stochastic environment. The deterministic analog of the problem was also solved. Then the two solutions were used to develop a measure of the value of a controlled environment and for the value of information about the stochastic environment.

*** 2.17 ***

Madden, J. Patrick. "Regenerative Agriculture: Beyond Organic and Sustainable Food Production." *The Farm and Food System in Transition*, no. 33. East Lansing: Michigan State University. Cooperative Extension Service, 1984.

Madden noted that the motive for studying regenerative agriculture stems from a growing concern about food safety, environmental quality, and the maintenance of agriculture on a long-term basis. New knowledge is needed to address these problems. The public is becoming increasingly concerned about how toxic chemicals in the environment and in foods and water affect our health. He concluded that "Researchers and extension personnel lack the knowledge necessary to create biologically and economically feasible alternatives to farmers' current dependence upon toxic, nonrenewable resources. A broader range of studies is needed to identify how a farmer can make the change from chemical-intensive farming to a farming system that is both less toxic and easier to sustain both ecologically and economically."

*** 2.18 ***

Martin, Marshall A., and Joseph Havlicek, Jr. "Some Welfare Implications of the Adoption of Mechanical Cotton Harvesters in the United States." *American Journal of Agricultural Economics* 59:4 (November 1977):739-744.

Martin and Havlicek conclude that "Both consumers and producers in the United States have benefited from the adoption of mechanical cotton harvesters. U.S. consumers received the larger portion of the economic welfare gains, approximately $230 million or about $1.25 per capita in 1947-49 dollars. The cost savings to the U.S. cotton producers as a result of the adoption of the harvesting technology was approximately 4.5 cents per pound of lint cotton (1.4 cents per pound of seed cotton). This cost savings is about 15 percent of the average real price received by farmers for lint cotton during 1952-69." The authors have noted, however, that the welfare gains from the adoption of mechanical cotton harvesters reported in this note were based on a partial equilibrium framework and did not reflect the direct impacts on the labor market. "Although the benefits to consumers and producers were substantial, the adoption of labor-saving technology contributed to the disequilibrium in the labor market in the South. Further research on this labor market disequilibrium should provide additional insights into the total magnitude of the social costs and benefits of the adoption of cotton harvesters."

*** 2.19 ***

Moffitt, L. Joe, Richard L. Farnsworth, Luis R. Zavaleta, and Marcos Kogan. "Economic Impact of Public Pest Information: Soybean Insect Forecasts in Illinois." *American Journal of Agricultural Economics* 68:1 (May 1986):274-279.

Previous agricultural pest management information studies have focused on the consequences of providing pest information to farmers who did not have access to other sources of pest information. In view of the proliferation of commercial scouting services, these authors used a preposterior decision model of pest management to examine the outcomes likely to follow from introduction of public pest information into an area with an existing commercial pest information market. They concluded that "Empirical estimates of the impact of a soybean insect forecast system in Illinois demonstrate that although farmers can benefit from public pest information, the implications for pesticide use depend on the reliability of public forecasts."

*** 2.20 ***

Peterson, Willis. "Land Quality and Prices." *American Journal of Agricultural Economics* 68:4 (November 1986):812-819.

Although land values reflect land quality, Willis concluded that "the quality characteristics that determine values are only partly related to agricultural uses. The results of estimating a reduced-form equation explaining differences in land prices among states suggest that nearly two-thirds of this variation is attributed to nonagricultural uses. Therefore, a land quality index based on raw land prices will be biased, which in turn will result in biased co-efficients in a production function or its dual." Willis constructed a cross-section land quality index, which excludes the influence of nonagricultural uses.

*** 2.21 ***

Schultz, Theodore W. "Tensions between Economics and Politics in Dealing with Agriculture." *1985 D.W. Brooks Lecture*. Athens: The University of Georgia, 1985.

In this broad-ranging, informative lecture, Schultz concluded that "Taking the long view, the critical component in economic modernization and development consists of human capital, that is, in the abilities and knowledge embodied in people rather than in the properties of land, other natural resources or other forms of physical capital. First and foremost over the decades are investment in population quality."

*** 2.22 ***

Shortle, James S., and James W. Dunn. "The Relative Efficiency of Agricultural Source Water Pollution Control Policies." *American Journal of Agricultural Economics* 68:3 (August 1986):668-677.

Shortle and Dunn examine the relative expected efficiency of four general strategies, which have been proposed for achieving agricultural nonpoint pollution abatement. Their emphasis is placed on the implications of differential information about the costs of changes in farm management practices, the impracticality of accurate direct monitoring, and the stochastic nature of nonpoint pollution. The possibility of using hydrological models to reduce, but not eliminate, the uncertainty about the magnitude of nonpoint loadings is incorporated into the analysis. Their principle result is that "appropriately specified management practice incentives should generally outperform estimated runoff standards, estimated runoff incentives, and management practice standards for reducing agricultural nonpoint pollution."

*** 2.23 ***

Sundquist, W. Burt. "Technology and Productivity Policies for the Future." *The Farm and Food System in Transition*, no. 4. East Lansing: Michigan State University. Cooperative Extension Service, 1983.

Sundquist notes that the U.S. farm and food system is now a high technology enterprise. He concludes that "The cost of research to 'maintain' this existing technology will increase in the future. And it will require even more new technology to provide for needed future growth in productivity. Effective public policies are needed to induce the development of this new technology....Where welfare effects cannot be adequately assessed, more general adjustment policies and programs are needed to correct for adverse effects."

*** 2.24 ***

Walker, David J. "A Damage Function to Evaluate Erosion Control Economics." *American Journal of Agricultural Economics* 64:4 (November 1982):690-698.

Walker developed an erosion damage function to measure on-site damage from agricultural soil loss. This method compared conventional farming and a conservation practice within a dynamic analysis treating conservation adoption year as a variable. Incremental damage from erosion, or marginal user cost was evaluated annually and included any cost to remedy lost future revenue from reduced yield. The damage function was applied to evaluate reduced tillage for wheat in the Idaho/Washington Palouse area. Walker concluded that "On shallower soils erosion damage provides conservation incentive, while on some deep soils erosion is economically rational." Properties of the damage function were explored through sensitivity analysis.

*** 2.25 ***

Zacharias, Thomas P., and Arthur H. Grube. "Integrated Pest Management Strategies for Approximately Optimal Control of Corn Rootworm and Soybean Cyst Nematode." *American Journal of Agricultural Economics* 79:3 (August 1986):704-715.

The authors used a dynamic programming model to determine approximately optimal management strategies for control of corn rootworm and soybean cyst nematode in Illinois. Decision alternatives for rootworm control included nontreatment, application of a soil insecticide, and rotation to soybeans. Alternatives for cyst nematode control were nontreatment, soil nematicide, resistant cultivars, and rotation to corn. State variables in the dynamic programming model were infestation levels of both pests, previous land use decisions, and expected product prices.

General Works

*** 2.26 ***

Haynes, Richard P., ed. *Agriculture and Human Values: Ethics and Values in Food Safety Regulation.* III:1 and 2 (1986):1-200.

The papers in this issue were written as part of a project designed by the Consumers' Union and supported by grants from the National Science Foundation and the National Endowment for the Humanities. As reported in an introduction to the study, the papers "represent an effort to repudiate an overly simplified picture of where scientific information and research fit into the picture of public policy.... The cases studied, therefore, though interesting and serious in their own right, are also useful as illustrative of the complexity of social policy issues in a centralized economic system which specializes various production functions and divorces them from consumption."

*** 2.27 ***

Horsfall, James G., et al. *Agricultural Production Efficiency.* Washington, DC: National Academy of Sciences, 1975. ISBN 0-309-02310-6.

In this product of a 15-member Committee on Agricultural Production Efficiency organized in May 1971, under the direction of the Governing Board of the National Research Council, the authors express the consensus of the National Academy of Sciences. The initial motivating concern arose out of long-term national statistics suggesting "possible faltering in our historically rising agricultural productivity." The food crisis of 1972 and the international energy crisis helped to dramatize food problems. The committee concluded that "the agricultural industry can continue to satisfy societal needs," but to do so "requires a fruitful and healthy cooperation among farmers, supporting industries, scientists, and educational leaders, in conjunction with the sustaining aid of helpful government policies and programs."

*** 2.28 ***
United States. Department of Agriculture and President's Council on Environmental Quality. *National Agricultural Lands Study, Final Report.* Washington, DC: U.S. Government Printing Office, 1981.

The crop land base identified in this study is composed of 540 million acres, of which 413 million acres are readily available crop land and 127 million acres are not being cropped, but have some potential for crops. In the late 1970s, when production-control programs were not restricting the land used for crops, about 370 million acres were being cropped. In comparison, as shown in reports of the U.S. Department of Agriculture, in the early 1920s about 365 million acres were in crops. The *National Agricultural Lands Study* concluded that growth of demand for land and the adequacy of supply depend on several factors. Growth of the demand for agricultural land will depend on U.S. population growth, expansion of exports of farm commodities, and the demand for land to grow crops for fuel. The adequacy of the supply of land will depend most on technological advance, which generally substitutes for land while increasing yields on land in crops.

CHAPTER 3

THE EVOLUTION IN FARM BUSINESS MANAGEMENT

Revolutionary changes in the financial, economic, and technological environment in which farmers operate has influenced and accompanied a revolution in farm business management. Prices are more volatile, risks are higher, capital gains and losses have occurred. Cash flow and tax management are part of the many concerns facing farmers.

"Farmers in turn have partaken in the gradual 'disenchantment' of agriculture--the replacement of intuition by calculation, and the progressive elimination of the mysteries of plant and animal husbandry by exposing them to scientific appraisal."

Howard Newby. *Agriculture in a Turbulent World Economy.*[1]

"Technological changes have brought increased specialization and dependence on capital and managerial skill."

"Resources, Food and the Future." *North Central Regional Extension Publication 222.*[2]

Revolutionary changes in the financial, economic, and technological environment in which agricultural producers operate are continuing to influence the accompanying revolution in farm business management. Increased price volatility for both the inputs farmers buy and the products they sell has resulted in significantly higher risk. Price changes in land and other inputs have created capital gains and losses that have overshadowed current income from the production of farm products. High interest rates on borrowed funds and increased use of debt capital in the larger scale farm businesses of the late twentieth century have combined to make financial inputs a significant part of farm business management.

The element of risk in farming has a new meaning. Farmers face not only the risk of variability of income due to changes in prices and yields, but they also face another crucial question--whether the farm operation can survive. For many farmers the focus of risk management is not on controlling or managing variability in income but on protecting the farm from failure, foreclosure, or bankruptcy.

Concerns about financial failure has stimulated a new emphasis on analysis of cash flows, liquidity of farm assets, and the collateral values that farmers can claim when seeking credit.

Tax management has also entered into farm business management. Taxation has been described as one of the "irritants of the real world" that has also become increasingly important in influencing decisions on the American farm. The tax treatment of inputs, capital purchases, and sale of products has directly influenced farmers' decisions on what inputs to buy, when to buy them, and the mix of labor and capital to use in farm production.

The rapid introduction of new technologies underlies many of the features of this changing environment. In recent years, new technology has been recognized as a major force in specialization of farm production, in the increase of capital investment in farm operations, and the development of special custom, business, and information services to farm operators.

It is important to understand how these evolutionary changes have influenced farm business management, and what they portend for the future of the American farm.

Types of Farm Business Organization

Changing costs of labor and capital and the specialization of farm production have influenced the type of farm business organization.

The pioneer farmer in America often worked along with his family and neighbors to provide the labor and make the decisions on how to operate the farm. Individual proprietorship is still the dominant form of farm business organization.

However as farms became larger, capital investments in land and machinery grew. Sometimes partnerships developed between a parent or parents and one or more of the children, or occasionally unrelated individuals, to share the costs of machinery, equipment, and land, as well as the labor needed for larger farm operations.

The corporation as a form of farm business organization came about for several reasons. In the West and South some publicly held industrial corporations entered into production of agricultural products. Much more frequently, however, as family farming operations grew and new family members entered the business, incorporation made possible an efficient way to divide ownership and distribute income and other benefits.

Probably the most important reason that many family farms have incorporated has been the federal income tax laws that provide tax savings for farmers and their families through incorporation. For example, in years when net incomes were favorable the taxable rate on income to individuals was higher than with corporations. However, with the lowering of individual tax rates under the 1986 Tax Reform Act, the benefits of incorporating to save taxes have been reduced. Certain expenses such as health insurance for employees could be counted as a business expense for the corporation but not for the individual farm operator. With corporate organization, it is also easier to transfer ownership from one generation to another when several family members are involved.

The 1982 Census of Agriculture showed that among 2,239,976 U.S. farms, 87 percent were individual operations, 10 percent were partnerships, and 3 percent were corporations. About one-half percent were other types of operations such as cooperatives, estates trusts, and institutional farms. Among the 59,792 corporations, 88 percent were family held and 12 percent were other than family held. Of the family-held corporations 97 percent had 10 or fewer shareholders and among the other-than-family-held corporations 84 percent had 10 or fewer shareholders.

The Beginnings of Farm Management as a Science

Although many statements about food production, animal husbandry, and improved farm practice may be found in ancient writings, they did not become part of a systematic plan for managing a farm. These early writings emphasized production techniques within individual farm enterprises.

The science of farm management, however, is a product of the twentieth century.[3] In 1910, the American Farm Management Association was organized, an organization that later merged with other professionals to become the American Farm Economics Association and then the American Agricultural Economics Association.

Farm management became a subject for college teaching and research and for extension programs for farmers. After World War I the Office of Farm Management was established in the U.S. Department of Agriculture. Later the Bureau of Agricultural Economics brought a renewal and an intensification of work in farm management and related agricultural economic subjects.

Farm management became an applied science in the United States due to highly developed transportation systems, the freedom to move commodities over a large area with variable natural adaptation in farm commodity production, the exhaustion of virgin soils suited to agricultural use, the development of a highly mechanized agriculture, the rise of research and extension programs to improve farming techniques, and the increasing influence of technology on output per farm worker. Competition among farm producers has tended to narrow profit margins, stimulate thought, and encourage study of the business aspects of farming.

Around 1890 many explanations and remedies were proposed for the unprofitable condition of agriculture. One view was that farmers should receive a cost of production price for their products. Scattered studies were made of the cost of producing the major farm products showing variations in costs among areas and among farmers of the same area.

Over many years, technological advances brought mechanical power and machinery for agriculture, improved varieties of crops and breeds of livestock, better soil treatment, methods of controlling animal and plant diseases, parasites, and insects, improved animal husbandry, better transportation and food processing facilities, and greater development in industries related to agriculture. While all these factors increased agricultural production, they created new problems by changing the economic environment in which farmers operated and by affecting the economic welfare of the individual farmer.

It was in the management of the farm as a whole that farm management emerged as a discipline and a profession. Farm management has become the science of correlating the many activities carried out on the farm. Farmers are faced with making managerial decisions whether or not they know all the technical details relating to crop or animal production.

Many of the early pioneers in farm management had been trained in crop production, animal husbandry, horticulture or other basic sciences.

One of the first educational and research efforts in farm management involved the use of model farms as examples for other farmers to follow. Later, surveys were taken. Farm management emerged as a blending of agricultural production with economics for the purpose of obtaining the largest profit, keeping the soil in a condition to attain maximum yields, and providing the farm operator with the largest measure of happiness.[4]

In one of the earliest text books in farm management published in 1911, Card wrote, "Executive ability and the proper adjustment of cog to cog in the business venture counts for more than soil fertility or intelligent crop management. To market a product advantageously is as essential as to produce it economically. Business methods are as important as production methods."[5]

From these early developments of farm management as a science has emerged more intensive applications of economic theory to the operation and decision making on the individual farm. However, the changing structure of American agriculture--the numbers and sizes of farms--has affected how different farms are managed and organized. Education, for the high school and college student as well as the adult operating farmer, underlies the entire revolution that has occurred in farm business management.

Farmer Participation and Involvement

Extension and research projects that involved farmers with surveys and model demonstration farms have led to farm business analysis and recordkeeping programs. Early in this century farm management specialists developed farm account books adapted to the farming operations of their respective states. Cooperative associations for this purpose evolved from these early efforts to encourage farm recordkeeping.

A fieldman supervisory and advisory role also evolved from these early experiences. Each fieldman services from 100 to 125 farmer clients, visiting them several times a year to counsel them on their recordkeeping and analyze their farm business. Some fieldmen also help in preparing federal and state income tax reports.

By 1986, farm business analysis associations were organized in Minnesota, Illinois, Kansas, Iowa, Wisconsin, Colorado, Kentucky, Nebraska, Alabama, and New Mexico. An estimated 17,600 farmers belonged to these associations and paid from $200 to $700 per year depending upon the size and type of their operations. They receive assistance from farm management fieldmen with recordkeeping and farm business analysis, and in some cases, income tax preparation. In other states, the Extension Service operates programs directly to assist farmers in recordkeeping and business analysis.

In the St. Paul Farm Credit District, about 14,000 farmers work with local Production Credit Associations to keep records under the Agrifax system. The St. Louis and Louisville farm credit districts have also operated farmer recordkeeping and farm business analysis programs.

Farm organizations and private companies have also developed recordkeeping and business analysis programs. A farmer participating in these programs may pay $500 to $1,500 per year for these services.

Professional Farm Management Services

As farm management became recognized as a scientific discipline as well as an important part of the knowledge base needed to make farm operations profitable, professional farm management services emerged. The first professional farm management service in the United States was established by D. Howard Doane in St. Louis, Missouri, in 1919. In 1929, The American Society of Farm Managers was organized and in 1937 the organization became the American Society of Farm Managers and Rural Appraisers.

For a fee, the professional farm manager represents the land owner in leasing and overseeing the rented land when the owner has other business or professional interests. Professional managers may also serve as consultants to farm owners or operators in organizing the farm business for maximum returns. Persons desiring to buy or sell farm land often seek help from rural appraisers to provide an evaluation of the fair market value.

Credit and Financial Analysis in the Farm Business

Farming has some unique credit requirements due to the nature of farm ownership and operation. On the traditional owner-operated farm, a new owner each generation must have the assets to buy the land or obtain the credit to do so. With the higher values of farm land and interest costs, and the increased capital requirements for machinery and equipment, the financing of a farming operation has become more difficult.

The operating credit used by the American farm has also increased--more than tripling from 1970 to the mid-1980s. Farm operations, being biologically based, have annual production cycles requiring credit to purchase the inputs early in the year with the expectation of salable products late in the year. With mechanization and modern technology, the cash expenses of farm operations have greatly increased, at least until recent years. At one time, farmers bought only a few inputs from off the farm. Farm receipts went toward purchase of family living items or a few new tools. By 1980, 93 percent of cash farm receipts were spent to buy items for farm production.

Since the founding of the federal land banks in 1916, a competitive credit system providing farmers with efficient, flexible, and adequate funding for farming operations has developed. The system includes commercial banks, the federally sponsored farmer-owned cooperative farm credit system, which lends money to farmers directly through the production credit associations, federal land banks, the government-owned Farmers Home Administration, and other private financial institutions and individuals.

At one time, the ability of a farmer to raise crops and livestock successfully was the measure of a successful farming operation. In more recent years, the ability to borrow funds has often been the key to success. Borrowing permits a farm operator to control productive resources, which in turn increases earnings and permits more rapid growth than if credit were not available. So the lender to the American farmer controls the allocation of credit and holds a key to the door to success for a farm operator. In choosing among potential borrowers, lenders have much to say about which farmers eventually succeed.

Numerous programs to help new farmers have focused on credit. The federal government, through the Farmers Home Administration, and a number of states have programs to help young farmers get established. The programs tend to emphasize low interest rate loans and liberal amounts of credit compared to the beginner's equity. A federal government guarantee on loans provided by private lenders has also been used.

Farm lenders often make their decisions to extend credit upon two measures of financial status: farm cash flow--the amount of cash income from farm and off-farm sources that is available to the farm household to make principal and interest payments, and provide for family living needs, and the debt-to-asset ratio--the ratio of total debt owned by the farm household to the total value of the assets it owns.

Part of the major financial stress faced by many farmers in recent years has been due to the declining value of their assets, mainly farm land, at a time when interest costs remained high along with other farm operating costs, and declines in farm prices are relative to costs of inputs.

High debt-to-asset ratios may not mean that a farm is in financial distress if the cash flow--the amount of income coming to the farm from sales of farm products and off-farm income-- meets the needs for family living and payments of interest and principal on the debt.

Consequently, managing farm debt and working out adequate cash flow has become part of the strategy for successful farm business management. Success in this area can be just as important as high yields or rapid gains for efficient production of crops and livestock.

Tax Management and Farm Management

Taxes and tax management affect the choices farmers make among varied marketing, financial, and production decisions. The tax reform bills in 1976, 1981, and 1986 each affected farmers' business decisions. Accountants, attorneys, and other advisers have established business and counselling services for farmers, landowners, and other investors to help them deal with various tax policies for their individual financial benefit.

The individual federal income tax is designed to impose a progressive tax on the individual's net income each year. But if gross income and its related expenses can be reported in different tax years, the amount of net income that an individual pays tax on during each year can be distorted.

Investments in farming that are taxed under preferential rules, such as the special income and estate tax rules for farmers, allow the creation of tax shelters. This tax shelter characteristic affects not only the total financial return from such assets but may also affect the pattern of ownership of such assets. The revenue losses to the U.S. Treasury due to the special farm tax rules were estimated at about $960 million in 1982. This loss is expected to be lower under the 1986 act.

The tax shelter aspects of tax law have encouraged the growth of individual farms. As long as there is other income that would be subject to tax, except for the tax shelter, taxpayers in higher tax brackets have more funds for growth and expansions than they would if the tax shelter did not exist.

Tax policy and the sheltering potential have affected the production and prices of farm commodities as well as management practices. The tax law provided incentives to develop orchard crops and expand dairy operations in spite of surpluses and excess production. Studies have linked favorable tax treatment to increased production and lower prices in the citrus and almond industries. Culling practices in the hog, beef, and dairy industries were altered in part to take advantage of the favorable taxation of capital gains and the cash accounting system affects the product selling and input purchasing decisions of farming.

The Tax Reform Act of 1986 attempted to reduce the tax shelter benefits for investment in agriculture. Although some sheltering is still possible, the limits written into the law are expected to reduce investments in agricultural enterprises for the purpose of sheltering funds from taxation.

Tax provisions in the past have affected the choice of a legal form of business. Frequently, lower overall tax rates could be achieved by incorporating the farm. Such lower tax rates provided more after-tax funds and a tax incentive to use them to expand the operation. Under the system of lower rates in the 1986 act, the tax benefits of incorporating will be reduced but family farming corporations may still be formed to provide a means of transferring land from one generation to the next and dividing and transferring ownership through shares of stock.

Because of the tax shelter potential, individuals with high incomes, in particular, have had strong incentives to report deductions as early as possible and delay reporting income as long as possible. Although the 1986 Tax Reform Act places some limits on tax shelters in agriculture, farm operators and farm owners will still watch the development of new tax rules carefully and take advantage of any remaining provisions that will make reduction of taxes possible.

The Computer Revolutionizes Farm Business Management

The evolution of the personal computer in the early 1980s provided an opportunity to apply modern business analysis methods to the farm by means of the latest technology. Mass production of the personal computer coincided with the financial problems on many farms. The poor farm economy may have slowed adoption of computers by some farmers, but for others it became an important tool in improving farm business management. By the mid-1980s, between 10 and 20 percent of the commercial farms owned and operated microcomputers.[6]

Agricultural software quality and quantity is improving steadily and the cost of computer hardware continues to fall. It is only a matter of time before most farmers will begin to use microcomputer technology.[7]

Because the total amount of data and information available to farmers is increasing as a result of new information technology, increased time for managerial activities can be expected. Telecommunications is becoming one of the farmer's most effective means of receiving information. By using a modem to connect his computer to a telephone line, farmers can bring in up-to-the-minute reports on commodity markets and weather. Marketing newsletters have become available over communication networks.

Causes and Effects of Advanced Management Technology

Improved farm technology has led to a substitution of capital for labor on American farms. Increased capital investment has stimulated the growth of individual farm operations to justify and pay for the increased capital. With larger farming operations has come a revolution in farm business management during the twentieth century.

The criteria for success in farming are no longer measured only by the crop yields or rates of production in livestock and poultry enterprises. The ability to use credit, schedule the needed cash flow to meet expenses, family living needs, and debt repayment, and retain enough money to meet all these needs has become the final measure of successful farming.

With this revolution in farm business management, farmers of the future will become more specialized. The work force will specialize in production of specific commodities, equipment management and operation, product marketing, accounting and allocation of funds for capital investments, operating purposes, and labor compensation. Consequently, farms of the future may be family operated, but several families are more likely to be engaged through partnerships or family farming corporations, just as other modern, competitive businesses operate.

The surviving farms of the future will place major emphasis on business management and proportionately less emphasis on the social orientation that views the single family farm as the only way to achieve happiness and success.

Endnotes

1. Howard Newby. "The Changing Structure of Agriculture and the Future of Rural Society." *Agriculture in a Turbulent World Economy.* Oxford: International Association of Agricultural Economists, Institute of Agricultural Economics, University of Oxford, 1986.

2. North Central Regional Extension Publication 222. Columbus: Ohio State University, 1984. p. 50.

3. H. C. M. Case and Donald B. Willliams. *Fifty Years of Farm Management.* Urbana: University of Illinois Press, 1957. p. 1.

4. Case and Williams. op.cit. p. 18.

5. Fred W. Card. *Farm Management.* Garden City, NY: Doubleday, Page & Company, 1911. p. v.

6. John Faulkner and Eric Brown. "PCs into Plowshares." *PC World* 4:10 (October 1986) pp. 222-231.

7. Arlin J. Brannstrom. "What's Available in Agricultural Software." *NCCI Quarterly* (North Central Computer Institute) 5:3 (September 1986) pp. 3-4.

Chapter 3: Annotated Bibliography

The Evolution in Farm Business Management

*** 3.1 ***
Guither, Harold D. *Heritage of Plenty, a Guide to the Economic History and Development of U.S. Agriculture.* Danville, IL: Interstate Publishers & Printers, 1972. LC 72-88713.

"Family farms of less than adequate size and those managed by ineffective operators will continue to be absorbed into larger units or to shift into part-time and retirement farms. Family farms will continue to dominate farming in the areas in which they are strong, but they will grow substantially in business stature," Guither concludes. The author provides a historic perspective on how U.S. agriculture developed from colonial times to the early 1970s. It contains many tables and charts of historic statistics on various facets of farming.

*** 3.2 ***
Lee, John. *Farm Sector Financial Problems, Another Perspective.* Agriculture Information Bulletin, no. 499. Washington, DC: Economic Research Service, U.S. Department of Agriculture, 1986.

Agriculture does not have an income problem today as much as it has a problem of absorbing large capital losses, Lee asserts. The farm crisis reflects the panic and pain associated with facing up to the reality of the loss. Although farming is still basically profitable, the huge debts incurred as a result of vigorous investment in the late 1970s undercut by declining land values in the 1980s have overwhelmed the debt-carrying capacity of earnings on some farms.

*** 3.3 ***
Lee, John. "Agriculture's Problems Require 'Macro' Solutions." *Farmline* 7:1 (December-January 1986): 4-7.

The 1985 Food Security Act is unlikely to solve some of the fundamental problems now facing American farmers. "General economic policies," Lee contends, "affect their broader interest as much as, or more than the farm bill." Interest on debt became the largest single cash cost of production in farming. The high cost of money has also helped to depress the value of farm assets. The higher the interest rate, the less a buyer can afford to pay for land.

*** 3.4 ***
Maunder, Allen, and Ulf Renborg, eds. *Agriculture in a Turbulent Economy.* Proceedings of the Nineteenth International Conference of Agricultural Economists, 26 August-4 September 1985. Oxford: University of Oxford, Institute of Agricultural Economics, 1986. ISBN 0-566-05225-3.

"Practical people...are impatient with the economists' tendency to use complicated ideas to tackle apparently simple problems. What may be called 'instant economics' has always appealed to the quick-witted layman...," commented Amartya Sen in his keynote lecture at this international conference. The increasing application of scientific and technological principles to the pursuit of profit in food production has produced a second agricultural revolution. In addition, state regulation of agriculture has profoundly altered both the structure of the industry and the day-to-day nature of life and work in the farm countryside. The changing structure of agriculture has become a topic of debate. In the more industrial countries, the differences between urban and rural society has changed dramatically. The nature and content of contemporary rural communities are quite different than strictly agricultural communities.

*** 3.5 ***
McElroy, Robert G. *Costs of Producing Major Crops, by State and Cropping Practice.* ERS Staff Report, no. AGES 860515. Washington, DC: Economic Research Service, U.S. Department of Agriculture, 1986.

Cost-of-production estimates are particularly important in extension and research programs. They provide farmers with a sense of whether they are achieving a level of efficiency and returns to maintain a viable farm operation. This report is a supplement to the annual Economic Indicators cost-of-production report. It describes the concepts and methods behind the estimates and how to interpret the costs and returns budgets.

*** 3.6 ***
Schlebecker, John T. *Whereby We Thrive, a History of American Farming, 1707-1972.* Ames: Iowa State University Press, 1975. ISBN 0-8138-0090-0.

"American farmers consistently strove to make money, so agricultural history also developed a form of economic history," Schlebecker explains. The first settlers hoped for wealth from a variety of enterprises, but they apparently expected little from agriculture except subsistence. The settlers and their backers in England had to modify both land laws and theory before a truly flourishing agriculture took shape. To further its colonial interests, the English government granted monopoly rights to certain companies for trade and settlement. Building on the success of its farmers, the United States after nearly 400 years emerged simultaneously as the world's leading industrial nation and one of the most urbanized. Free enterprise and freedom of choice, long honored in the United States, seemed an adequate explanation.

*** 3.7 ***
United States. Department of Agriculture. *Contours of Change.* The Yearbook of Agriculture, 1970. Washington, DC: U.S. Government Printing Office. LC 79-609792.

In the first section of this valuable reference, 19 authors provide details on "The Agricultural Revolution." Behind such fascinating titles as "Systems Come, Traditions Go," "King Cotton Blasts Off," and "What's Happened to Farming?" are the details about people, farming, conservation, and environment and how they interact but continue to provide abundant production of food and fiber commodities in the face of change.

*** 3.8 ***
Van Chantfort, Eric. "Farmland Values: Where's the Bottom?" *Farmline* (May 1986):4-6.

Farm land values dropped in 37 states in 1985. Nationally, the average decline amounted to about 12 percent. The U.S. Department of Agriculture economists believe the declines should slow down, but a sudden turnaround doesn't seem likely.

Types of Farm Business Organization

*** 3.9 ***
McDonald, Thomas, and George Coffman. *Fewer, Larger U.S. Farms by Year 2000--and Some Consequences.* Agriculture Information Bulletin, no. 439. Washington, DC: Economics and Statistics Service, U.S. Department of Agriculture, 1980.

The number of U.S. farms will likely decline by about a third in the next 20 years, if present trends continue, while the number of large farms with annual sales of more than $100,000 will quadruple. Family farms will continue to dominate, but the influence of small farms will wane. More large farms will probably mean more farm corporations, more specialization in what farms produce, agricultural production concentrated among relatively few farms, and fewer young people getting started in farming because of the high capital requirements. The uncertainty of energy, its cost and availability, could have opposite effects on growth of farms, the authors conclude.

*** 3.10 ***
Miller, Thomas A., Gordon E. Rodewald, and Robert G. McElroy. *Economies of Size in U.S. Field Crop Farming.* Agricultural Economic Report, no. 472. Washington, DC: Economics and Statistics Service, U.S. Department of Agriculture, 1981.

"Increasing farm size does not necessarily increase farm efficiency or productivity," the authors conclude. They studied farms producing wheat, feed grains, and cotton in seven field crop regions to determine the importance of size in affecting costs of production. They found that as farm size increases, per unit costs decline at first and then are relatively constant over a wide range of sizes. Net farm income on small farms was limited by low volume of production not by high costs.

*** 3.11 ***
Owen, Wyn F., ed. *American Agriculture, The Changing Structure.* Lexington, MA: D.C. Heath and Company, 1969. LC 74-10655.

"Much has been contributed by economists who have sought to open up the pandora's box of 'technological progress,'" Owen declares. In the United States, agriculture is undergoing• a transformation similar to the enclosure movement that accompanied the Industrial Revolution in England. Over time one organizational structure is replaced by another. The limits to farm size in the United

States hinge more upon what may be termed spatial and temporal limits than the social and political considerations stressed by Thomas Jefferson. However, some authors still believe that since the contemporary structure of agriculture is a creation of government, so too, government will be instrumental in the molding of the future system.

*** 3.12 ***
United States. Department of Agriculture. *A Time to Choose: Summary Report on the Structure of Agriculture*. Washington, DC: U.S. Government Printing Office, 1981.

"As a farmer, I had always felt there had to be a better way to make farm policy...After six years in Congress, I was absolutely convinced," Bob Bergland declares in the foreword to this report. Based on a series of dialogue hearings between the Secretary of Agriculture and interested persons at locations across the country during 1979, the report describes American agriculture and its environment and lists areas of concern that affect the future structure of agriculture. These include landownership, soil and water conservation, tax policy, commodity, credit, public research and extension, agricultural labor, and trade policies.

*** 3.13 ***
University of Nebraska. Department of Agricultural Economics. *Corporation Farming, What Are the Issues?* Report, no. 53. Lincoln, NE: 1969.

At a time when the traditional family farm is threatened, concerns about large scale farming often turn to corporations and the implications for this type of farm business organization. The major issues in these workshop proceedings centered on the rapid expansion in size of farms requiring large amounts of capital, the possible competitive advantages that large farms including corporations may have, the consequences of proposed restrictive legislation on corporate farming, the adjustment problems faced by rural people excluded from the modern industrialized economy, and the economic and social costs and benefits of an industrialized agriculture. The consensus seemed to be that the goals of public policy, which deal with the issues emerging from the industrialization of agriculture, should be to manage the rate of change, and to encourage resource adjustment for maximum human welfare rather than to freeze human and other resources in place.

The Beginnings of Farm Management as a Science

*** 3.14 ***
Boehlje, Michael D., and Vernon R. Eidman. *Farm Management*. New York: John Wiley & Sons, 1984. ISBN 0471046884. LC 83-23589.

"No longer can the successful farmer have primarily a production and technology orientation; he or she must understand and skillfully apply management concepts in the areas of marketing and financial management as well," Boehlje and Eidman assert. Management involves implementing those choices and plans once a decision is made and then monitoring actual performance and comparing that performance to planned expectations. This integrated approach to farm management is the book's main focus.

*** 3.15 ***
Calkins, Peter H., and Dennis D. DiPietre. *Farm Business Management, Successful Decisions in a Changing Environment*. New York: Macmillan Publishing Co., 1983. ISBN 0023183209.

The period from the early 1980s to 2000 will see tremendous change in the environment in which agricultural producers must operate. Laws to limit agricultural pollution and soil erosion, the movement to reduce energy use, changes in dietary preferences of Americans, will all require that producers have a clear conception of the essential principles of farm management science, the authors emphasize. Students are provided with a plan to face the real world of risk, needs for budgeting, and acquisition of capital to carry out agricultural production.

*** 3.16 ***
Card, Fred W. *Farm Management*. Garden City, NY: Doubleday, Page & Company, 1911.

"Higher crop and animal production does not represent all there is to good farming," the author cautions in his introduction to this pioneer work in farm management. His prophetic conclusion: agricultural teaching and agricultural practice will both give greater heed to the business management of the farm in the years to come than in those gone by.

*** 3.17 ***
Case, H. C. M., and Donald B. Williams. *Fifty Years of Farm Management*. Urbana: University of Illinois Press, 1957. LC 56-6707.

Case and Williams reveal the historic roots of farm management as a science and indicate who the principal actors have been. They have not written a book on techniques and application, but rather an account of the academic development of teaching, research, and extension work.

*** 3.18 ***
Haworth, Paul Leland. *George Washington, Farmer*. Indianapolis: The Bobbs-Merrill Company, 1915. LC 15-19355.

In 1788 Washington wrote to a prominent English author and scientific farmer, "the more I am acquainted with agricultural affairs, I can no where find so great satisfaction in those innocent and useful pursuits." As revealed in his extensive diaries, he placed major emphasis on improving crop varieties and the quality of his livestock. However, Washington also believed that records were essential to successful farming. Not only did he record his agricultural and other experiments, but he also detailed accounts of the weather so he could decide the best time for planting various crops. He kept exact accounts of financial dealings and a special book for the accounts of the estate of his stepchildren.

*** 3.19 ***

Kay, Ronald D. *Farm Management: Planning, Control, and Implementation.* 2d ed. New York: McGraw-Hill Book Company, 1986. ISBN 0070334943. LC 85-18203.

Kay designed his textbook for students taking their first course in farm and ranch management. It is organized around three basic functions of management: planning, implementation, and control. Good luck or bad luck cannot explain all the differences observed in the profitability of farms and ranches even among those that have about the same amount of land and capital available. The difference between profitable and unprofitable farms is due to management, defined as the judicious use of means to accomplish an end. The second edition reflects changes in income tax regulations and includes a discussion of microcomputers in agriculture.

Farmer Participation and Involvement

*** 3.20 ***

Illinois Farm Business Farm Management Association. *FBFM, the First 50 Years.* Urbana, IL: The Association, 1974.

"As agriculture has changed from a pastoral way of life to a complex industry fraught with regulations and many societal demands, the need for farm business records in the decision-making process is even greater than before," the authors assert. They relate 50 years of evolution of an institution that has set an example for other states to follow in organizing associations for farm business analysis. By far the largest cooperative farm business analysis service in any state is the Illinois Farm Business Farm Management Service with about 7,500 farmer members serviced by 60 fieldmen in 10 associations across the state. The service began in November 1924 when 52 farmers signed agreements in Woodford County. Earlier experience by extension advisers in these counties had shown that more farmer cooperators would complete their records when extension advisers made personal contact to supervise farmers in keeping their records.

*** 3.21 ***

Wilken, Delmar F. "Educational-Service Units: Farm Business-Analysis and Recordkeeping Program." *Paul A. Funk Recognition Program, Papers by the Recipients of Awards in 1984.* Urbana: University of Illinois. College of Agriculture, 1984.

Farm business analysis and recordkeeping grew out of an extension and research project conducted by the Department of Agricultural Economics at the University of Illinois. The farms in the Midwest and Great Plains are largely operated as sole proprietor businesses and have more unpaid family workers than paid laborers. The author sees such a program as a means to enhance farm living, demonstrating that farm records verify a job well done and provide standards of performance and financial results in both extension and research activities. This management revolution has been made possible by education and research. Most successful farmers have invested more in their human capital than did their fathers and grandfathers who farmed before them. Farm operators' participation has been achieved through improved recordkeeping and farm business analysis, support of cooperative farm business associations, enrollment with private business analysis services, and adoption of the latest accounting and analytical technology such as computers.

Professional Farm Management Services

*** 3.22 ***

American Society of Farm Managers and Rural Appraisers. *Official Directory of General Membership.* Denver, CO: The Society, 1987.

In its 1987 directory, the American Society of Farm Managers and Rural Appraisers listed 3,700 members in 46 states, Canada and several foreign countries. The society has seven membership classifications and twenty-three working committees. There are 37 state associations or chapters. Every year since 1949, the society has sponsored intensive education programs in rural appraising, rural condemnation appraising, and farm management. The society awards professional designations of Accredited Farm Manager (AFM) and Accredited Rural Appraiser (ARA) to professional and academic members who have met the designated high experience, education, ethical, and performance standards in their respective fields.

*** 3.23 ***

DeBraal, J. Peter, and Gene Wunderlich, eds. *Rents and Rental Practices in U.S. Agriculture.* Proceedings of a Workshop on Agricultural Rents. Oak Brook, IL: Farm Foundation and Economic Research Service, U.S. Department of Agriculture, 1983.

With the decline in land values and financial stress among farm operations, the question of rents and leasing practices becomes a critical issue. Workshop contributors discuss share contracts and risk, legal aspects of leasing, determining cost of land by observing rents, future of farm land leasing, and current leasing practices. Historically, many American farms were and are worked by tenants. The extent and nature of farm tenancy varies across regions and crops, as well as over time, although share tenancy is the predominant type.

Credit and Financial Analysis

*** 3.24 ***

Barry, Peter J. *Financial Stress in Agriculture: Policy and Financial Consequences for Farmers.* AE-4621. Urbana: Department of Agricultural Economics, University of Illinois, n.d.

The author reports the findings of a study analyzing the financial consequences of various policy options and strategies that farmers might follow in responding to the financial stresses of the 1980s. A clear implication of the results of the study is the need to target the development and use of public assistance programs to the financial characteristics of the intended recipients. Farmers with relatively low leverage can survive without assistance. Those with 70 percent debt-to-asset ratios are only maintained in their situation while those with 40 percent debt-to-asset ratio benefit the most from public assistance programs.

*** 3.25 ***

Brake, John R. "Financing Agriculture in the Future." *The Farm and Food System in Transition*, no. 10. East Lansing: Michigan State University Cooperative Extension Service, 1984.

The author briefly relates the sources of credit for farmers, how de-regulation affects competition in financial markets, and the effects of government farm programs on farm asset values. He relates how credit is a major determinant of farming success, and the proper role of government in providing credit to agriculture.

*** 3.26 ***

Cooke, Stephen C., and Ronald D. Knutson. "Is Bigger Better: Economies of Size in Agriculture." *Policy Choices for a Changing Agriculture, North Central Regional Extension Publication*, no. 66. Columbus, OH: Ohio State University, 1987.

About 300,000 of the largest farms produce 70 percent of total output and account for over 95 percent of the total net farm income. The smallest, 60 percent of all farms, produce only about 6 percent of all output and have negative net farm incomes. These numbers increasingly reveal a two-tier structure of large and small farms. An important force determining the number and size of farms has been economies of size.

*** 3.27 ***

Harrington, David H., Donn A. Reimund, Kenneth H. Baum, and R. Neal Peterson. *U.S. Farming in the Early 1980s, Production and Financial Structure.* Agricultural Economic Report, no. 504. Washington, DC: Economic Research Service, U.S. Department of Agriculture, 1983.

"No longer is farming insulated from developments in the rest of the Nation and the world. New technology has contributed to these changes...," the authors conclude. They present an excellent perspective on the numbers and sizes of farms by income classes, regional shares of commodity receipts, the extent of contract farming and vertical integration, and profiles of the types of farm business organization.

*** 3.28 ***

Kadlec, John E. *Farm Management, Decisions, Operation, Control.* Englewood Cliffs, NJ: Prentice-Hall, 1985. ISBN 013305053. LC 84-3315.

We have entered a new era in agriculture--the management era, Kadlec declares. The management era involves: (1) greater emphasis on continual analysis for key decisions and adjustment of the farm business to changes in technology, market conditions, laws, and regulations; (2) increased importance of gaining control of adequate capital; (3) greater need to control all aspects of the farm business; (4) increased market orientation; (5) more emphasis on hiring and supervising workers and dealing with people; (6) increased importance of business legal organization and tax management; and (7) continued emphasis on efficiency of crop and livestock production.

*** 3.29 ***

Penson, John B., Jr., Danny A. Klinefelter, and David A. Lins. *Farm Investment and Financial Analysis.* Englewood Cliffs, NJ: Prentice-Hall, 1982. ISBN 0133050378. LC 81-15715.

Farmers and ranchers need to consider many things when making investment and financing decisions. The authors provide worksheets incorporating the necessary steps rather than complex formulas. Topics covered include financial statement analysis, evaluating investment opportunities, and financing new business investments. The authors direct their book to farmers and ranchers, both young and old, who want to develop an operation of a more profitable size. Lenders and agribusinessmen should find it useful since farmers may ask them for financial advice.

*** 3.30 ***

United States. Economic Research Service. *Financial Characteristics of U.S. Farms, January 1986.* Agriculture Information Bulletin, no. 500. Washington, DC: U.S. Department of Agriculture, 1986.

"Lower commodity prices, lower farm exports, and declining values for farm land have taken their toll on farm businesses," this report concludes. Farmers' cash flow positions improved in 1985 largely because of high direct government payments and increased Commodity Credit Corporation loans. However, farmers' equity positions continued to erode because of the crop in real estate values. The proportion of farms in the most serious financial difficulty (high debt-to-asset ratios and negative cash flows) declined from 12.5 percent in 1984 to 11 percent in 1985. The Corn Belt, the Northern plains, and the Lake States accounted for over 60 percent of these farms. Financial stress was highly concentrated among farms with annual sales of $40,000 or more and among grain and general livestock farms.

*** 3.31 ***

United States. Economic Research Service. *Economic Indicators of the Farm Sector, National Financial Summary 1985.* Washington, DC: U.S. Government Printing Office, 1986.

Net cash income in 1985 was estimated at $44 billion, significantly above the narrow range of $33-$39 billion where it had been since 1978. For the first time in over a decade, farmers reduced their cash expenses by slashing outlays across nearly all expense categories. This report details the farm sector's financial status on 1985 earnings, assets, and liabilities. Total farm assets (including farm households) declined nine percent to $866.8 billion as of 31 December 1985, mainly because of a continued decline in farm real estate values. Total farm production expenses declined four percent to $136.1 billion for the first year-to-year decline in over 30 years, except for a slight drop in 1983 that was a direct result of the payment-in-kind program, which cut acreages sharply.

*** 3.32 ***

United States. Department of Agriculture. *Cutting Energy Costs.* The Yearbook of Agriculture, 1980. Washington, DC: U.S. Government Printing Office. LC 80-600168//r82.

Prepared at a time when energy costs were a major concern, this volume illustrates another part of modern farm business management, the efforts to cut production costs. The authors provide valuable insights about how farmers have struggled to hold down their production energy requirements. A new revolution may be in progress--one in which agriculture's own renewable energy supplies may be used increasingly to fuel farm machinery, heat farm buildings, dry grain, and serve many other purposes.

Tax Management and Farm Management

*** 3.33 ***

Boehlje, Michael. "Taxes and Future Food and Fiber System Structure and Performance." *The Farm and Food System in Transition*, no. 25. East Lansing: Michigan State University. Cooperative Extension Service, 1984.

In this brief paper, the author describes the personal and corporate income, labor and employment, property, sales, excise, and estate taxes and explains how they affect decisions of farm and agribusiness managers. He also describes tax shelters and how they affect farming operations. He concludes that tax policy has encouraged expansion of farm operations and capital investments, thus encouraging substitution of capital for labor.

*** 3.34 ***

Breimyer, Harold F., and James Tucker. *How Federal Income Tax Rules Affect Ownership and Control of Farming.* North Central Regional Extension Publication 37. Special Publication, no. 32. Urbana: University of Illinois Cooperative Extension Service, 1974.

The authors point out that capital investment in agriculture is filled with income tax considerations. Any part of agriculture, and any class of investors, that can best exploit tax advantages thereby gain a competitive edge over other parts and other investors because they can outbid for resources. Special income tax rules applicable to agriculture permit farmers to mismatch income and costs, thereby reducing their tax liabilities. Such provisions include: (1) use of cash accounting, (2) the immediate deduction of some capital expenses, and until the 1986 Act became effective, (3) the capital gains treatment for income from assets for which costs may have been deducted as a current expense. Since cash accounting ignores inventories, a farmer can deduct costs of production and control the tax year in which income is realized by storage of crops and careful timing of sales. Tax rules that affect the management of agricultural enterprises offer the possibility of inducing management practices that are financially attractive but not basically economic, they conclude.

*** 3.35 ***

Carman, Hoy F., and Sermin D. Hardesty. "Federal Income Tax Policies and Financial Stress in Agriculture." *Agricultural Finance Review* 47 (Special Issue, 1987).

"There is growing evidence that short run gains from provisions for cash accounting, current deduction of capital expenses, capital gains income, accelerated depreciation and investment tax credits may adversely affect longer-run agricultural returns," Carman and Hardesty conclude. Special income tax rules that apply to agricultural investments have been extensively used and exploited both by farmers and nonfarm investors for more than 25 years. The impacts include structural changes as well as the fact that investment behavior influenced by income tax rules was characteristic of many farm firms currently facing the greatest financial difficulties.

*** 3.36 ***

LeBlanc, Michael, and James Hrubovcak. "The Effects of Tax Policy on Aggregate Agricultural Investment." *American Journal of Agricultural Economics* 68:4 (November 1986):767-777.

Tax policies are effective in promoting agricultural investment; nearly 20 percent of net investment in agricultural equipment during the period 1956-78 is attributed to tax policy. From 1956-78, tax policy has stimulated, in real dollars, over $5 billion in net investment in equipment and in excess of $1 billion in structures. Tax policy has affected the composition of the capital stock with the investment tax credit biasing investment in favor of equipment rather than structures.

*** 3.37 ***

Musser, Wesley N., Bernard V. Tew, and Fred C. White. "Choice of Depreciation Methods for Farm Firms." *American Journal of Agricultural Economics* 68:4 (November 1986):980-989.

Accelerated depreciation methods are usually considered to increase the present value of after-tax cash flows for farm firms compared to straight line methods. A review of the theoretical foundations of this conclusion indicated that they require an assumption of constant marginal income tax rates, which is inappropriate for many farm firms. Straight line methods are preferred with income tax rates below maximum levels and/or lower discount rates. General recommendations on depreciation methods are therefore impossible.

*** 3.38 ***

Saxowsky, David M., David L. Watt, and W. Allan Tinsley. *Tax Implications of Liquidating a Farm Operation*. Washington, DC: Extension Service, U.S. Department of Agriculture, 1986.

There are possible tax consequences of disposing of farm business assets. These tax effects can, if not considered and planned for, be significant for owners of farm businesses, both in and out of financial difficulty. A tax obligation may result from gains from disposal of business assets for more than their tax basis, from use of the alternative minimum tax, from recapture of investment tax credits, from taxation of forgiven debt, from cancellation of a contract for deed, or from recalculation of estate taxes.

*** 3.39 ***

Uchtmann, D. L., and J. T. Cross. "The Effects of Tax Policy on the Structure of Agriculture: Tax-Induced Substitution of Capital for Labor." *The American Journal of Tax Policy* 3 (Spring 1984): 115-152.

The authors separated the economic factors and the government-induced factors that would encourage farmers to substitute capital for labor. Then they identified the tax policies that encouraged farmers to substitute capital for labor. These included accelerated depreciation, investment tax credit, social security, unemployment insurance, and some combined effects.

The Computer Revolutionizes Farm Business Management

*** 3.40 ***

Brannstrom, A. J. "What's Available in Agricultural Software." *NCCI Quarterly* 5:3 (1986):3-4.

Substantial libraries of easy-to-read reference books have been written to assist farmers and agricultural students to understand the computer and its applications to farm management and farm business analysis. In 1986, the North Central Computer Institute at the University of Wisconsin identified 1,100 commercial and 700 university authored software programs for agricultural use. These programs were most frequently written for financial management and crop and livestock production. Available software includes financial management programs for accounting, linear programming systems, payroll systems, tax management, financial analysis, budgets and cash flow, marketing, land purchase, lease and sale, and labor scheduling. Livestock management programs include nutrition and ration formulation, herd management, beef feedlot management, beef cow and calf, dairy herd, swine, and poultry management. Crop management programs involve specific crops such as corn, soybeans, cotton, hay, fruits and vegetables, nursery, and forestry. Programs have also been developed for use of chemicals and their application, irrigation, planting, harvesting, weather analysis, and conservation practices.

*** 3.41 ***
Cardiff, John. *Farming and the Computer*. Seattle: Group Four Publications, 1985. LC 85-9880.

While the computer has long been a useful addition to modern industry, the potential in agriculture has been exploited with marked success in the 1980s. Cardiff deals with the computer system, the potential uses of a computer on the farm, where to shop, and how to house the computer on the farm. He has designed a vehicle by which farmers can consider the possibility of using personal computers as a farm business management tool, it being based on six years of study, observation and evaluation by the author, and additional experience by various contributors.

*** 3.42 ***
Erickson, Duane E., Royce A. Hinton, and Ronald D. Szoke. *Microcomputers on the Farm: Getting Started*. Ames: Iowa State University Press, 1985. ISBN 0813811562. LC 85-11951.

The authors present the essentials of acquiring and using a small computer in agricultural operations. The computer may become a necessity for farmers to remain successful. In buying a computer and software, the authors advise looking at the functions required in the farm operation, user friendliness, vendor support, and adequacy of documentation. Computers, like other machinery and equipment, require a capital investment and the time and skills to use the investment profitably.

*** 3.43 ***
Faulkner, John, and Eric Brown. "PCs into Plowshares." *PC World* 4:10 (October 1986):222-223.

While the personal computer may not make farming a fail safe venture, it can provide better information with which to gamble, the authors of this article on application of computers to farming operations conclude. They observe that to remain successful, farmers will have to learn the computer just as their great grandfathers had to learn to use the combine and the corn picker. Larger farm operations can afford to use the computer, as well as other advanced technologies that require major capital investments, and profit from its use more effectively than small farm businesses. Others have reported that in troubled times, bankers look for financial responsibility and good recordkeeping. PC accounting systems can show off records in impressive style. In future years, the key role for personal computers will be to provide accounting and help the farmer be accountable to the bank. The computer may not make farming a fail-safe venture, but it can provide better information on which to make the traditional high risk decisions in farming.

*** 3.44 ***
Hein, Norlin A., and Mark Wilsdorf. *A Computer for Your Farm, Some Things to Think About*. North Central Regional Extension Publication, no. 247. Columbia: Cooperative Extension Service, University of Missouri, 1987.

The authors define a computer as a "cross between a calculator and a file cabinet." The major arguments for using a computer are its usefulness for calculation and information filing, its speed in calculations and storing information, and its assistance in gleaning management information. This guide gives seven steps to buying a computer, choices of hardware and software, and a glossary of terms for those who want to understand what computers are all about.

*** 3.45 ***
Purdue University. School of Agriculture. *On-Farm Computer Use*. Conference proceedings. W. Lafayette, IN: Purdue University, 1983.

This 350-page volume is an encyclopedia on computers, computer services for recordkeeping, analysis, and farm financial management. The conference session included sessions for the beginner considering what to buy, for the advanced computer user and the programmer, plus specific applications for crop, machinery, pork, and dairy production. The theme of this volume is best expressed by one speaker who declared, "If you expect to prosper, nay survive, in production agriculture, you will have to become as knowledgeable about computerized record-keeping as you are about crop and animal production."

*** 3.46 ***
Rasmussen, William O., C. T. K. Ching, Lucille A. Linden, Patricia A. Myer, V. Philip Rasmussen, Jr., Roy S. Rauschkolb, and Charlotte B. Travieso. *Computer Applications in Agriculture*. Boulder, CO: Westview Press, 1985. ISBN 0813300614, 0813300622 (paperback). LC 85-8855.

Since U.S. agriculture is the most technologically advanced in the world, it is only natural that agriculture should be linked with today's focal point of high technology--the computer, the authors assert. The authors provide a brief introduction to computers in general, describe the way computers are currently used in agriculture, and provide a comprehensive list of references to enable readers to keep abreast of the rapidly changing field of computer applications in agriculture and a glossary of computer terms.

*** 3.47 ***
Sistler, Fred E. *The Farm Computer*. Reston, VA: Reston Publishing Co., 1984. ISBN 0835918629, 0835918610 (paperback). LC 83-13937.

The author has designed this book to help

people in agricultural related fields understand the use and application of computers to the farm or farm related business. Computers can do many useful things and they do not have to be intimidating. While they cannot do your thinking for you, they can make your work, time, and decision making abilities more productive. Sample programs are included as a basis to develop your own programs.

*** 3.48 ***

Sonka, Steven T. *Computers in Farming, Selection and Use*. New York: McGraw-Hill Book Company, 1983. ISBN 0070596530 (paperback). LC 83-7911.

The ability to acquire and use information is one of the keys to successfully managing today's farm or ranch. As the amount of information needed increases, the effort each manager devotes to collecting and analyzing that information also grows. This book documents how computers can assist agricultural producers acquire and manage information. The overriding purpose is to provide the background necessary so that its readers can use the computer technology to make better agricultural decisions.

Causes and Effects of Advanced Management Technology

*** 3.49 ***

United States. Senate. Committee on Agriculture, Nutrition, and Forestry. *Farm Structure, a Historical Perspective on Changes in the Number and Size of Farms*. 96th Congress, 2nd Session. Washington, DC: U.S. Government Printing Office, 1980.

During the late 1970s when the structure of farming became a major public issue, the Senate Committee on Agriculture, Nutrition, and Forestry published this collection of papers dealing with various perspectives on the numbers and sizes of farms from a historical, commodity, and rural community perspective. Authors include agricultural economists from the Land Grant Universities across the country and the U.S. Department of Agriculture.

General Works

*** 3.50 ***

Richardson, William B., William G. Camp, and William G. McVay. *Managing the Farm and Ranch*. Reston, VA: Reston Publishing Company, 1982. ISBN 0-8359-4236-8.

The authors have written a survey of farm management in clear, simple language with no prerequisite training in economics or mathematics required. While it may be considered a traditional text in farm management, it provides questions at the end of each chapter, which allow the reader to apply concepts and receive another perspective on the science of farm management.

CHAPTER 4

THE EVOLUTION IN MARKETS AND MARKETING OF AGRICULTURAL PRODUCTS

A principal problem in agriculture has been the difficulty of coordinating production with market needs. Lack of change in the way individual farmers market their products is often blamed for oversupply, lower prices, and total failure of a farming operation.

"Marketing is part and parcel of the modern productive process, the part at the end that gives point and purpose to all that has gone before."

U.S. Department of Agriculture. *Yearbook*, 1954.

"A principal problem in agriculture has been the difficulty of coordinating production with market needs."

U.S. Congress. Office of Technology Assessment, 1986.

Producing and marketing are two distinct functions that, while in the past have been viewed as two components of farms, have usually been viewed together as two major components of a modern business or industry.

Through the years, farmers have adopted new production technology and raised their productivity. However, the lack of change by individual farm operators in the way they market their products is often blamed for oversupply, lower prices, and total failure of a farming operation. This chapter reviews the evolution in markets and marketing and considers the relationship of this evolution to problems of the American farm.

Markets and Marketing in Historic Perspective

When pioneer families settled and established the first farms, each farm family tried to produce what they could to meet their own needs for food, clothing, and shelter. Further processing of farm commodities for consumption usually occurred on the farm. Tasks such as butchering livestock and preserving meat, growing and canning fruits and vegetables, shearing sheep and processing the wool were all part of the activities of the pioneer farm family or household. In those times before our modern industrial system was established, each farm family had to be as self-sufficient as possible.

Producing for home use came first. Then any surplus production was traded locally for items that could not be produced on the farm.

As farms expanded, new technology boosted production and transportation of farm products became possible. As buyers and sellers met at designated locations, markets provided an outlet for selling the farm commodities of a community or a region. Development of markets was made possible by constructing roads, canals, and railroads to move the products to points where they could be assembled and sold.

In the 20th century the development of a national system of paved highways and modern truck transportation has accelerated movement of farm commodities from producer to consumer. However, the high costs associated with assembly market trading has led many farmers and buyers to develop other ways of doing business. Direct marketing has emerged as a substitute for assembly markets. In 1940, about 75 percent of all livestock were handled in assembly markets and only 25 percent by direct marketing. By 1960, the assembly market share of all livestock sold by farmers had declined to 50 percent. By 1980, it had declined to less than 20 percent.

Marketing in a Modern Perspective

Marketing of farm commodities is the process of moving a commodity from the farm to the consumer. The goods that farmers grow and sell must be stored, transported, processed, and delivered in the form and at the time and places that consumers desire.

The storing, transporting, processing, and delivery may be considered the simpler parts of marketing. Today, specialization is the general rule. Fewer and fewer farmers sell directly to consumers. Rather, each function that occurs in marketing, between the time a farm commodity is first offered for sale and the final purchase, is performed by agencies or persons who have some particular advantage or skill.

The modern marketing system has several functions:

1) To move the desired varieties of farm and food products to consumers in the desired forms and conditions at the lowest possible cost.
2) To provide a living for people working in the marketing system and to yield reasonable returns to the capital and management skills devoted to it.
3) To find and develop new markets in the sense of moving new products, better products, or more of the old products either at home or abroad.

Marketing is the performance of business activities that direct the flow of goods and services from the producer to the consumer in order to reach consumers at the time, place, and in the form they desire at a price they are willing to pay. The people and firms involved in moving farm commodities from the farm to the final consumer are often referred to as "middlemen," as they are in the middle between the farmer and the final consumer.

Agricultural marketing in the 1980s has some unique characteristics compared to nonagricultural products. Farm output is largely raw material that requires further processing before it can be used by the ultimate consumer. Livestock must be converted to meat, wheat into flour and bakery products, grains into cereals, and milk into various processed dairy products.

Quality as well as the total production of commodities varies from year to year and from season to season. Some years the quality is high and in others it is low.

Growth of Marketing

Marketing of farm products has developed in importance and complexity as specialization of activities has increasingly separated producers from potential consumers. In 1790 when 90 percent of the population of the United States were farmers, marketing was relatively simple since about 90 percent of the population produced their own food and the remaining 10 percent were close by.

With the industrial revolution, manufacturing of various food, clothing, and shelter materials moved from small cottage industries into larger factories. Specialization also influenced the growth of cities. With the disappearance of the necessity for people to produce all of their basic needs, they could leave the land and work in larger groups. With the growth of cities came enlarged marketing efforts to serve the needs of people living and working some distance away.

Throughout most of our early history, one of the pressing agricultural marketing problems was providing adequate transportation facilities at a reasonable cost. Interest developed in having government help the marketing system to function smoothly and efficiently.

Turnpikes, canals, railroads, highways, and airports have been built to improve our transportation system, which carries all kinds of farm commodities from producer to consumer.

A complex and costly marketing system is necessary when a broad mass market is to be served. The principal concept of the American marketing system is that of consumer direction. The production and marketing of food and other farm-produced commodities is in response to the consumers preferences as registered by direct purchases of those products.

MARKETING ISSUES AND CONCERNS

Efficiency

Efficiency is a major concern of those who study and observe modern agricultural marketing systems. They try to determine if a system is too complicated, if the technological progress in marketing is keeping pace with the advances in farming, and if it is possible to reduce waste, overlapping, and duplication in marketing and services.

Research is carried out to gain vision, foresight, and knowledge and to reduce marketing costs, waste, and spoilage. Much research is directed toward increasing the productivity of labor because labor cost constitutes so large a part of the total marketing bill. As marketing firms grow larger, improvement in management methods is also needed to facilitate efficient marketing.

Cost

Payments for many different marketing services such as assembling, storing, transporting, processing, wholesaling, and retailing are all contained in the total marketing bill.

In 1981, U.S. farmers received $85 billion for the commodities that went to American consumers or were exported. The job of marketing cost $200 billion. The different functions were: processing, $76.2 billion; transportation, $15.1 billion; wholesaling, $23.4 billion; retailing, $38.0 billion; and food service $47.1 billion.

Marketing farm commodities has some major differences when compared with marketing industrial goods. Most farm commodities are bulky and production is widely dispersed and a considerable distance from consumers. There are long fixed lags between the decision to produce and the actual output, which differs with annual grain crops such as wheat or corn, perennial crops such as apples or oranges, or livestock products like meat, milk, or eggs. Most farm crops mature once a year and they cannot be produced continually. Many are highly perishable such as milk and fresh fruit. The system must deal with widely varying production from one year to another and must accommodate these variations by storage and other means.

Setting Prices

When the large public assembly markets operated at major transportation centers, large numbers of buyers and sellers were present to negotiate prices based on the supply and demand on that day. With the decline in public assembly markets and the growth of direct marketing, some are concerned that prices may not be established under competitive conditions and that buyers may have an advantage over sellers.

Commodity markets in Chicago, Minneapolis, Kansas City, and New York provide the opportunity to buy or sell futures contracts for specified amounts and grades of commodities. The Chicago Board of Trade is the largest commodities futures market in which wheat, corn, oats, soybeans, soybean oil, and soybean meal are sold every working day. The Chicago Mercantile Exchange operates a futures contract market for live hogs, feeder and fed cattle, and pork bellies. Some producers are concerned that the futures markets do not reflect the proper value of the commodities, but such markets do provide an opportunity for buyers and sellers from all over the world to buy and sell these commodities.

Grades and Standards

A grade is an established measure of quality that is applied during the grading of the commodity in question. The purpose of grading is to establish a common language, understood and used by buyers and sellers, as a basis of judging the quality of a commodity in relation to its sale price. The quality of grain, cotton, eggs, butter, potatoes, onions, and most other farm products are designated by federal grades.

The grading of grain became a major issue in the late 1980s as questions were being asked about the adequacy of grades and standards being used. Grain grades, established in 1916, were being studied because the grading standards were not rewarding farmers and grain marketing firms for producing and marketing the highest grades.

Fair Treatment in the Marketplace

One of the persistent and overriding goals of farmers, seemingly unachievable at times, is to secure equitable treatment and avoid unfairness in marketing their products. Federal agencies involved in helping to assure fair treatment in marketing farm commodities include the Packers & Stockyards Administration, the Federal Grain Inspection Service, and the Agricultural Marketing Service. Milk marketed in the major urban markets is sold under federal marketing orders. Fruits and vegetables may be sold under federal marketing orders and agreements to control quantity and quality and assure stable prices. The United States Warehouse Act provides a measure of protection for farmers delivering crops for storage to public warehouses and to facilitate financing arrangements for stored grain and other farm commodities.

Vertical Coordination in Marketing Farm Commodities

Vertical integration or market coordination is the process of directing and harmonizing the inter-related and sequential decisions and processes involved in producing and marketing farm commodities. This process has developed as part of the revolution in the marketing of farm commodities primarily during the last half of the 20th century. The decision about what to produce, where, how, and when to sell are coordinated from farm to final distribution point.

Broiler production has been the most dramatic example of expansion in contract production since World War II. This development involved the favorable combination of technological development in breeding, feeding, disease control, and economic innovation.

The broiler industry has been cited as a model of the organization that may characterize most of U.S. farming in the future. The growing of broilers by independent producers selling through an open market has been largely replaced by a highly concentrated, integrated, and industrialized business with sectors for on-farm production, processing, and distribution. The location of broiler production and processing shifted and total output increased fivefold from the end of World War II to the early 1980s.

Vertical coordination varies widely among the various crops. For instance, the most vertically integrated and contract crops are sugar beets with 100 percent, vegetables for processing with 98 percent, and seed crops with 90 percent. Less than 8 percent of the feed grains and 9 percent of the food grains are produced under contract or are vertically integrated.[1]

From the producer's standpoint, production contracts increase price stability and availability of operating capital. Processors may provide capital for production under the contract terms.

With contract production, the processing vegetable growers may not be able to sell their production unless they have a contract. So the opportunity to grow processing vegetables is almost completely limited to those with contracts. Contracting thus has become a barrier to entry for new growers since processors tend to stay with established producers with known performance. Processors also favor larger producers since the cost of contract supervision is lower for a few large producers than more numerous small producers.

Arguments can be made for and against vertical integration. On the positive side, it has contributed to the uniform size and quality of farm products. It has contributed to increased efficiency and reduced costs. On the negative side, potentially adverse consequences may result. Contract integration radically changes the role of the traditional, independent farmer. The farmer may lose control of the commodities grown under a production integrated arrangement.

The Role of Farmer Cooperatives

A cooperative is a business firm owned and controlled by those who use its services. Decisions are made and control exercised by the owners as patrons rather than as investors. The basic purpose is to hold costs to a minimum consistent with the quality of services demanded by the patron-owners. Through cooperatives, farmers have extended their business activities significantly beyond the farm.

Public policy in the future can have a dramatic impact on the direction of farmer cooperative development. Public policy has supported an owner-operator system for agricultural production. Cooperatives are a hand-in-glove economic extension of that system. Changes in the extent and kind of government regulations and support programs could drastically alter the nature of agriculture and the role of cooperatives.

The Capper Volstead Act provides both limited anti-trust protection for farmers and consumer protection from price exploitation. The Secretary of Agriculture can take action against farmers' cooperatives that use their market power to unduly enhance price.

The American Farm in an International Market

International trade for U.S. farm commodities has become increasingly important. The growth of agricultural exports in the 1970s provided expanded markets for the products of many American farms. By 1980, the output from one out of every three harvested acres in the U.S. was sold abroad. The decline in agricultural exports in the 1980s brought declining prices, more government programs to reduce output, and financial losses for many American farmers.

U.S. farmers need the additional demand that international markets represent. The American farm is capital intensive. It is more efficient when high capital costs are spread over more acres. The vast expanses of fertile land in the midwestern and western states allow those efficiencies to be achieved. If one-third of acreage were not devoted to exports and farmers were constrained to operate at only two-thirds of capacity, their per acre costs of production for interest, fertilizer, and other inputs would be higher and their competitive position with farmers in other countries would be less.

In the early 1980s, agricultural exports accounted for about one of every four jobs in agriculture. Toward the end of the decade, agricultural leaders and government policy makers were wondering how they could develop new markets for farm commodities and regain some of the foreign market share lost from earlier years.

Growth in World Trade

The trends in food production around the world and the growth in trade has produced a world food system. This is a new concept that is not well understood. In 1960, about 90 percent of all food produced was consumed in the country where it was produced. By the early 1980s, only 85 percent was consumed in the country where it was produced. World trade in food commodities has grown much faster than total production.

While the need and the demand for food will continue to grow well past the end of the 20th century, it will probably not grow as rapidly as during the 1970s. As a major producer and consumer of agricultural products with one of the world's most efficient agricultural production systems and the potential to greatly expand production in the future, the United States can continue to be a major force in the global food system. What role it will play depends largely on the policy choices that are made.

Alternative Marketing Strategies

The United States has three alternative marketing strategies to stimulate farm commodity exports: free trade, market development, and trade agreements. The argument for a free trade has met with only limited success because of the failure to persuade other countries to eliminate trade barriers and trade distorting mechanisms such as export subsidies.

Free trade has also been hampered by world economic recession and by high domestic farm price support policy. The decision to lower price supports following passage of the 1985 Food Security Act was an effort to improve the competitive position of the American farm in world markets. However, recovery from the decline in farm exports that occurred in the early 1980s will be slow and the high level of export growth in the 1970s should be considered a special situation rather than a long term trend.

Policies Related to Trade

If the United States chooses to maintain high, rigid price supports to augment farm income and protect inefficient producers, it will likely price itself out of world markets and necessitate setting up trade barriers to insulate U.S. producers from the world market. With high price supports, either programs to restrict production or programs to subsidize exports will be required. Under such conditions, the United States will probably not be a major force in the global food system.

The American farm can participate fully in expanding markets for the commodities when it can produce at a comparatively lower cost. Under such conditions, both farmers and consumers will be exposed to the variability of world markets. Some smaller and less efficient farmers will be forced out of business, but those who remain will benefit from growth in world markets and be less dependent on government price and income supports.

Marketing and Public Policy Issues

A major concern among many people is how well the marketing system is working, how much government regulation is needed and how it should be administered.

Prices and pricing are frequently at the head of the list of issues. Price fluctuations that may be substantial from day to day raise questions about whether the marketing system is fair to all producers.

Market Access

The availability of only one or two market outlets raises questions about access to the market. The problem of low volume markets with a lack of competition among bidders, inadequate information, and high potential for price manipulation has concerned many farmers as producers have moved away from centralized markets to country markets. Electronic markets can provide a mechanism to centralize the price negotiation process for farm commodities without the physical assembly of buyers, sellers, and products at one location. Such a system should reflect more accurately the overall marketwide conditions, access to markets should be enhanced, and total marketing costs should be reduced. Although electronic marketing would not reduce price uncertainty nor enhance farmers' market power, electronic technology would be used to revitalize the open market system.

Price Risk and Forward Pricing

By using contracts, farmers can achieve more certainty about what price they will receive by setting the price in advance of harvest, or even before it is produced. However, the quantity to be delivered, the quality, and the timing of delivery remain uncertain.

The approval of commodity-options trading on the major futures markets illustrates further government action to help producers reduce risk from wide price fluctuations. However, use of options requires that farmers who use them understand how to use options to reduce price changes and the costs they must pay for risk reduction.

Public Reporting of Market Information

With more direct marketing, vertical integration, and contract agreements, the amount of public information about prices has declined. Some marketing economists feel that public reporting of prices paid for agricultural products should be required so that producers could make more informed decisions when marketing their crops and livestock.

Exclusive Agency Bargaining

Farmers have often faced frustration in marketing since they could not predict future prices or match the power of large buyers. A means for farmers to join bargaining associations, just as workers have joined labor unions, and bargain for prices under contract with processors of farm commodities suggests a way to solve the pricing problems. Farmers who want to achieve exclusive agency bargaining face some major efforts to secure the legislation to accomplish this goal.

Expanding Marketing Orders

Marketing order programs were initiated to bring orderly conditions to chaotic markets in milk, fruits, and vegetables. Expanded use of marketing orders may provide producers a means of providing new marketing activities or coverage for additional commodities. Though potentially able to improve producer incomes in the short run and provide a more orderly marketing, costs to producers, consumers, and marketing firms must be considered.

For producers, marketing order programs as have been authorized and operated in the mid-1980s do not limit production and thus have limited ability to materially enhance prices. They can be and are used to create more orderly marketing and to stabilize prices. For consumers, the primary benefit of marketing orders is to assure continuing uniform and stable supplies over a long period. Such stabilization could increase prices in the beginning, but over a period of time prices may not be increased.

Marketing Boards

These are marketing institutions long established in Australia, New Zealand, Canada, and the United Kingdom. Marketing boards have been proposed in the U.S. as a substitute for private grain exporting. Marketing boards have generally been established following periods of depressed and uncertain agricultural prices and producer discontent with the structure and functioning of the agricultural marketing system. Usually legislation is required to establish a marketing board and its membership, functions, and powers.

Preparing for the Future

The way Americans live, work, and eat has changed since World War II. More meals are eaten alone. An estimated 16 percent of dinners and 41 percent of lunches are eaten away from home. New lifestyles and new information about linkages between diet, health, and longevity have changed consumers' preferences for various types of food. Changing preferences, rising incomes, and changing relative prices have resulted in consumers demanding more variety and convenience, fewer calories, less animal fat, more lean protein, and more fruits and vegetables.

The demand for food is changing, influenced by income and population growth, age structure of the population, household size, mobility, labor force participation, access to information, and health expectations. Consumer preferences are changing the mix of foods bought in the domestic market. Farmers can no longer assume that all food produced is desired by consumers or that they have the capacity to eat the quantities of food being supplied. These changes will affect production decisions, farm prices, incomes, structure of agriculture, and marketing practices, especially for those producers who are dependent on the domestic market.

In the 1970s, rising exports contributed to farm prosperity; in the 1980s, falling exports contributed to farm despair. The future contributions of exports to farm prosperity depend on whether domestic policies affecting the American farm can be structured to support rather than discourage exports, whether domestic and foreign trade barriers can be reduced, whether a role for assisting development of other countries leads to expanding markets for U.S. farm products, and whether U.S. producers can produce competitively with producers in other countries.

Endnotes

1. Alden C. Manchester. "The Farm and Food System, Major Characteristics and Trends." *The Farm and Food System in Transition*, no. 1. East Lansing: Michigan State University. Cooperative Extension Service, 1983.

2. T. Kelley White. "The Global Food System & the Future U.S. Farm and Food System." *The Farm and Food System in Transition*, no. 20. East Lansing: Michigan State University. Cooperative Extension Service, 1984.

Chapter 4: Annotated Bibliography

Markets and Marketing in Historic Perspective

*** 4.1 ***
Henderson, Dennis L. "Farm Product Assembly Markets in the Future Farm and Food System." *The Farm and Food System in Transition*, no. 12. East Lansing: Michigan State University. Cooperative Extension Service, 1983.

Public assembly markets were once the major connecting points for trading in most agricultural commodities. These included local and country auction markets as well as central wholesale markets located in cities having major railroad terminals such as Chicago, St. Paul, and Omaha. Central terminal markets were first organized around the time of the Civil War as railroads became the major means of transporting farm commodities "back east" where most consumers lived. Many of these markets still persist, particularly for livestock, fruits, and vegetables. Country markets first emerged as a means for assembling a large enough volume from local farmers for trainload shipment to a distant terminal market. Later as truck transportation became feasible for hauling farm commodities and as processing plants moved out of the big cities closer to their sources of supply, country markets evolved into direct buying points for processors. These markets have two distinguishing characteristics: assembly and public trading. The commodities to be traded are physically assembled in the market facility at the time of the sale, and trading is generally open to anyone who wishes to buy or sell.

*** 4.2 ***
Manchester, Alden C. "The Farm and Food System, Major Characteristics and Trends." *The Farm and Food System in Transition*, no. 1. East Lansing: Michigan State University. Cooperative Extension Service, 1983.

Many different industries provide inputs both to farming and to the processing and distribution of farm products. Vertical coordination is particularly important in the food industry because of the distances from farm to final distribution point, the large number of specialized firms involved, the uncertainty of prices, supplies and qualities of products, and the urgency of marketing perishable products. The vertical coordination may be carried out in two ways: by prices, or by other means such as contracts, to communicate the needs of each firm to others in the marketing system. The shift to vertical coordination has shifted the decisions of what and how much to produce from the farmer to the food marketing firms. Also there is a trend toward use of administered payments such as contracts or agreements with bargaining associations. Some farm commodities have undergone more vertical integration than others. A United States Department of Agriculture study in 1980 estimated that about 96 percent of the fluid grade milk, 99 percent of the broilers, 99 percent of the eggs, and 90 percent of the turkeys were produced under vertical integration and production-marketing contracts. Only 3 percent of the hogs and 16 percent of the fed cattle were vertically integrated or under production contracts at that time.

*** 4.3 ***
Marion, Bruce, and NC-117 Committee. *The Organization and Performance of the U.S. Food System*. Lexington, MA: Lexington Books, 1986. ISBN 0669112208. LC 85-45106.

The authors examine all facets of the food marketing system from the farm to the supermarket. The major objectives were: (1) to describe the structural characteristics of industries involved in the food system and identify the changes and causes of change, (2) to describe the vertical organization, systems of price discovery and coordinating mechanisms of selected commodity markets, (3) to describe the legal environment of the food system, and (4) to identify and evaluate the consequences of alternative public and private actions that could be taken to alter the future organization, control, and performance of the food system.

*** 4.4 ***

Stanton, B. F. "What Forces Shape the Farm and Food System?" *The Farm and Food System in Transition*, no. 2. East Lansing: Michigan State University. Cooperative Extension Service, 1983.

Important forces have helped to create our complex, interdependent farm and food system and are now bringing about still further change. What has changed is the time and effort spent in performing the production, processing, and marketing functions by which we obtain our daily food supply. Forces that have helped change and shape the farm and food system are new technology, growth of markets, a public transportation system, and an industrialized and service-oriented society.

*** 4.5 ***

United States. Department of Agriculture. *Marketing*, 1954 Yearbook of Agriculture. Washington, DC: U.S. Government Printing Office, 1954.

Secretary of Agriculture Benson declares that greater emphasis than ever before has been placed on marketing as a mainspring of our national lives. Marketing involves competition, tension, and differences of opinion. The contributors to this volume provide a wealth of information on the bases for marketing, sales off farms, central markets, food retailing, trade abroad, storage, transportation, processing, grades, standards, cooperatives, fair dealing, prices, pricing, and efficiency.

Marketing in a Modern Perspective

*** 4.6 ***

Chafin, Donald D., and Paul H. Hoepner. "Improved Profits with Continuous Marketing." *Journal of the American Society of Farm Managers and Rural Appraisers* 50:1 (April 1986):40-42.

The concept of 730-day marketing offers farmers a superior, new method for capturing market profits. Continuous marketing is a drastic change from speculative, onetime pricing. Two innovative procedures differ markedly from most marketing strategies. First, the ability to speculate without holding grain in bins allows the producer to capture both a strong basis and a high price level. Second, pricing decisions are not an irreversible, onetime decision, but a constant process. A farmer holds a futures market position all the time, either short the new crop or long the old crop.

*** 4.7 ***

Collins, Robert A., and Roy D. Nelson. "The Inter-Year Effect of Routine Marketing on Farm Income and Utility." *The North Central Journal of Agricultural Economics* 8:2 (July 1986):257-268.

Evaluations of marketing strategies commonly include simple comparisons of hedging alternatives. These comparisons usually omit statistical tests for differences among means and variances. This paper presents evidence that hedging can cause meaningful reductions in income variability without significantly reducing the mean. Despite this possible risk reduction, however, application of excess return to variability methodology shows that routinely applied hedging rules of thumb yield outcomes that exhibit no discernible difference in expected utility. This calls in question the value of disseminating advice to farmers based on simple mean-variance comparisons.

*** 4.8 ***

Elitzak, Howard. "Where the Food Dollar Goes." *National Food Review* 35 (Fall 1986):23-26.

The farm value of the $343.6 billion, which consumers spent for foods originating on U.S. farms in 1985, represented about a fourth of the total compared to about a third in 1975. The marketing bill accounted for the rest. The smaller share from a few years earlier is due both to weak farm prices and rising marketing costs.

*** 4.9 ***

Kilmer, Richard L., and Walter J. Armbruster, eds. *Economic Efficiency in Agricultural and Food Marketing*. Ames: Iowa State University Press, 1987. ISBN 0070072418. LC 82-18016.

Authors of papers presented at a symposium analyzed the current state of agricultural marketing and food marketing efficiency. The underlying concepts of economic efficiency and the latest methodological developments and their applications were discussed. References provide a comprehensive review of recent literature.

*** 4.10 ***

Kinsey, Jean, and Tom Cox. "Consumer Demand for Agricultural Products in the United States: A Moving Target." *Policy Choices for Changing Agriculture*. Columbus, OH: National Public Policy Education Committee, 1987.

The way Americans live, work, and eat has changed over the past 30 years. Changes in domestic food consumption patterns affect agricultural producers, processors, consumers, and taxpayers. Changes in income and relative prices, as well as demographic, social, and educational trends influence consumption patterns of food and fiber. Consumer preferences for convenience, variety, fewer calories, less animal fat, lean protein, and more fruits and vegetables are changing the mix of foods bought in the domestic market. Farmers cannot assume that all food produced is desired by the consuming public or that consumers have the capacity to eat the quantities of food being supplied.

*** 4.11 ***

Leath, Mack N., and Lowell D. Hill. *Grain Movements, Transportation Requirements, and Trends in United States Grain Marketing Patterns during the 1970s*. North Central Research Publication, no. 268. Urbana, IL: 1983.

"Recent trends suggest that volumes of grain handled by Gulf and Pacific port elevators will continue to grow as exports expand, while the volumes handled by Great Lakes and Atlantic port elevators will be maintained at current levels," the authors conclude. This report of a regional research project concludes that the marketing of grain in the U.S. involves complex interregional movements. It requires a large transportation capacity. The marketing patterns for individual grains and the transportation requirements of all grain movements are presented.

*** 4.12 ***

Smith, Edward G., Ronald D. Knutson, and James W. Richardson. "Input and Marketing Economies: Impact on Structural Change in Cotton Farming on the Texas High Plains." *American Journal of Agricultural Economics* 68:3 (August 1986):716-720.

The authors identify the degree of pecuniary economies existing for input purchases and marketing on cotton farms in the Texas Southern High Plains. The results indicate that substantial cost and marketing economies are being realized by the largest farms in the region.

*** 4.13 ***

United States. Economic Research Service. *Food Costs...from Farm to Retail*. Washington, DC: U.S. Department of Agriculture, 1987.

The farm value share is what the farmer receives from the dollar the consumer spends for foods in retail foodstores. Over time, the share reflects relative changes in farm and retail food prices. Consumers, farmers, and legislators want to know what causes food prices to change and the difference between what farmers get for the commodities they sell and how much consumers pay for them. That difference in economic terms is called the farm to retail price spread. To answer this question, the U.S. Department of Agriculture measures price spreads for foods originating on farms. In 1986, the farm value share of the retail price was 62 percent for eggs, 54 percent for beef, 49 percent for milk, 46 percent for pork, and 55 percent for chicken. For canned tomatoes, the farm value share was only nine percent, and for white bread only 7 percent because of the high processing costs for these commodities in comparison with farm costs.

Growth of Marketing

*** 4.14 ***

Polopolus, Leo. "Agricultural Economics beyond the Farm Gate." *American Journal of Agricultural Economics* 64:5 (December 1982):803-810.

The industrialization of American agriculture has diminished farm numbers and the farm population. While total agricultural production has increased, the input, finance, service, processing, transportation, and distribution industries surrounding production agriculture have now become the dominant economic components of the food and agriculture system. In terms of value added, employment, or other measures of economic importance, the food and fiber system beyond the farm gate is roughly twice that of production agriculture. Society will receive significant net benefits from increased public and private investment in research and education relating to the food system beyond the farm gate, the author asserts.

*** 4.15 ***

Thurman, Walter N. "The Poultry Market: Demand Stability and Industry Structure." *American Journal of Agricultural Economics* 69:1 (February 1987): 30-37.

The stability of the demand for poultry meat and the specification issues in its estimation are explored. Thurman concludes that the demand for poultry meat shifted out in the early 1970s. At the same time, the demand relationship between poultry and pork changed from substitution to independence. He also concludes that poultry price is predetermined by cost of production while quantity is determined by demand.

Marketing Issues and Concerns

*** 4.16 ***

Farris, Paul L. "Concentration Policy for the Farm and Food System." *The Farm and Food System in Transition*, no. 45. East Lansing: Michigan State University. Cooperative Extension Service, 1985.

In many farm and food system markets, competitive forces are strong and these forces influence the way resources are allocated and incomes are distributed. However, in some markets, impediments are apparently suppressing competitive forces and channeling economic activity away from goals that many people would consider socially desirable. Through the years a considerable divergence of opinion has developed concerning policies used to achieve performance goals. One point of view is that competition requires competitors, and it works best when the number of competitors is sufficiently large so that they impose mutual restraints on each other.

*** 4.17 ***

Kauffman, Daniel E., and James D. Shaffer. "Forward Contract Markets: Can They Improve Coordination of Food Supply and Demand?" *The Farm and Food System in Transition*, no. 40. East Lansing: Michigan State University. Cooperative Extension Service, 1985.

The authors explore the problem of coordinating supply and demand under the real world conditions of uncertainty and examine the possible use of contract markets. The major benefit of the contracting system would be the improved coordination of supply and demand. This will happen only if contracts cover a significant portion of production. Going beyond the small amount of contracting that is currently being done in most agricultural products will be difficult. A system of forward contract markets could contribute to improved coordination of supply and demand in the U.S. food system and allow the greatest individual freedom of action.

*** 4.18 ***

United States. Senate. Committee on Agriculture and Forestry. *Marketing Alternatives for Agriculture, Is There a Better Way?* Com. Print, 94th Congress, 2nd Session, April 7, 1976, Washington, DC: U.S. Government Printing Office, 1976.

The American agricultural producer and the marketing system that delivers his products to consumers across the Nation and around the world are testimonials to modern technology and organization. Any modification of the marketing system must be done with care and awareness of the impacts. A large share of the increased costs of food and fiber is due to cost factors between the farmer's gate and the consumer's kitchen. A distressing fact is that while farm prices go down as well as up, the cost of the marketing functions only go up.

Vertical Coordination in Marketing Farm Commodities

*** 4.19 ***

Mighell, Ronald L., and William S. Hoofnagle. *Contract Production and Vertical Integration in Farming, 1960 and 1970*. ERS-479. Washington, DC: Economic Research Service, U.S. Department of Agriculture, 1972.

In this brief publication, the authors offered the first comprehensive effort to measure the extent of contract production and vertical integration in farming. Although a technical distinction is made between contract farming and vertical integration, the point being made is that freely competitive markets are reduced when larger proportions of a commodity go to market by way of contracts or integration.

*** 4.20 ***

Reimund, Donn A., J. Rod Martin, and Charles V. Moore. *Structural Change in Agriculture, the Experience for Broilers, Fed Cattle and Processing Vegetables*. Technical Bulletin, no. 1648. Washington, DC: Economics and Statistics Service, U.S. Department of Agriculture, 1981.

The authors provide unique insights into the modernization of raising broilers, feeding cattle, and growing vegetables for processing, and its influence upon structural change in the production of these commodities. They identify the stages of change and describe their consequences.

The Role of Farmer Cooperatives

*** 4.21 ***

Caves, Richard E., and Bruce C. Petersen. "Cooperatives' Tax 'Advantages': Growth, Retained Earnings, and Equity Rotation." *American Journal of Agricultural Economics* 68:2 (May 1986):207-213.

Cooperatives are subject to full tax integration for the bulk of their income, while corporations' net income is subject to what is known as a classical form of taxation. This paper derives the condition under which full tax integration gives the cooperative a lower cost of equity capital and develops a model to examine the effect of taxation, together with equity rotation, on the growth path of cooperatives. Cooperatives, under current financial practices, are capable of extremely high short-term growth rates, but they are not sustainable.

*** 4.22 ***

Centner, Terence J. "Agricultural Cooperatives: Retained Patronage Dividends and the Federal Securities Acts." *North Central Journal of Agricultural Economics* 6:1 (January 1984):36-47.

The investment of retained patronage dividends in agricultural cooperatives has been accompanied by equity redemption programs that fail to provide for the timely return of patron's interests. The unfairness of some equity redemption programs with regards to deceased and retired cooperative patrons could result in judicial or legislative action. Recent judicial interpretations of the Federal Securities Acts suggest that dissatisfied cooperative patrons could commence legal actions under these acts for damages, injunctions, and other appropriate relief.

*** 4.23 ***

Knoeber, Charles R., and David L. Baumer. "Understanding Retained Patronage Refunds in Agricultural Cooperatives." *American Journal of Agricultural Economics* 65:1 (February 1983):1-37.

Most equity capital in agricultural cooperatives is raised by retaining a share of member patronage refunds. The collective decision to retain patronage refunds reflects the desire of the median member. Variables affecting the share of patronage refunds retained are expected rates of return on cooperative equity capital and on farming assets, their variances, their covariance, and expected future share of patronage and its variance. The behavior of major regional supply cooperatives was found to be consistent with the model developed by the authors.

*** 4.24 ***

Royer, Jeffrey S., and David W. Cobia. "Measuring the Equity Redemption Performance of Farmer Cooperatives." *North Central Journal of Agricultural Economics* 6:1 (January 1984):105-112.

Cooperatives need to adopt more effective systems for ensuring that equity is held by patrons in proportion to use. A disparity index was developed to measure the degree to which it is not. To compare the simulated performance of alternative equity formation and redemption programs, disparity index values were calculated for a hypothetical cooperative. Results demonstrate that plan performance depends largely on patron histories of patronage and equity investment. Nevertheless, revolving fund plans with revolving periods up to five years usually had the lowest disparity index values. Generally, plans with the lowest disparity index values required the greatest annual patron investments.

*** 4.25 ***

Sexton, Richard J. "The Formation of Cooperatives: A Game-Theoretic Approach with Implications for Cooperative Finance, Decision Making, and Stability." *American Journal of Agricultural Economics* 68:2 (May 1986):214-225.

Sexton departs from the traditional organization-oriented approach to cooperative analysis. He exploits the cooperative's functional similarity to vertical integration to examine individuals' incentives to form cooperatives. A model of formation of a purchasing cooperative is presented and developed as a game with the core of a solution concept. Core existence is examined for both single- and multiple-cooperative configurations, and cooperative finance methods are examined relative to finding core-compatible allocation rules. The results provide insight into a cooperative's equilibrium output, stability, decision making, financing methods, and choice of open or restricted membership.

*** 4.26 ***

Wills, Robert L. "Evaluating Price Enhancement by Processing Cooperatives." *American Journal of Agricultural Economics* 67:2 (May 1985):183-192.

Section two of the Capper-Volstead Act requires the secretary of agriculture to determine whether a cooperative has unduly enhanced its prices. There has been no consensus about what level of price enhancement is too high. Wills contrasts pricing of cooperative and proprietary brands of differentiated food products. Empirical evidence indicates that market share and advertising do not generally give cooperatives any more power to enhance prices than they give proprietary firms. The author suggests a standard for undue price enhancement, the predicted price level of proprietary brands in similarly structured markets.

The American Farm in an International Market

*** 4.27 ***

Christiansen, Martin F., ed. *Speaking of Trade, Its Effect on Agriculture*. Special Report, no. 72. St. Paul: Agricultural Extension Service, University of Minnesota, 1978.

Foreign sales have become basic to U.S. farmers' economic health and to the future growth of U.S. agriculture. This publication provides an economic perspective on trade as it affects agriculture. Its chapters include a historical review, basic concepts of trade, relationships between domestic agricultural and trade policies, international trade arrangements, quantitative dimensions, and current and emerging issues. The glossary defines many terms not readily understood.

*** 4.28 ***

Dunmore, John C. "Competitiveness and Comparative Advantage of U.S. Agriculture." *Increasing Understanding of Public Problems and Policies--1986*. Oak Brook, IL: Farm Foundation, 1987.

The U.S. farm sector is highly dependent on sales to foreign markets to fully utilize its productive capacity. In the early 1970s, the U.S. farm sector had 293 million acres in production, with substantial acres idled. That year, 71 million or 24 percent were used for production of exports. By 1980, 352 million acres were in production, an increase of nearly 60 million acres. The expanding export market was responsible for most of the additional resources that were drawn into U.S. agriculture in the 1970s, and in fact, bid some resources away from production for domestic use. While declines in export value, volume, and market share are evidence of a loss of competitiveness, they provide little insight into comparative advantage.

*** 4.29 ***

Ek, Carl W. *Grain Quality: Background and Selected Issues*. Report, no. 85-229 ENR. Washington, DC:

Congressional Research Service, The Library of Congress. Washington, DC: U.S. Government Printing Office, 1985.

Ek discusses several facets of the current debate over the quality of U.S. grain exports. Farmers and members of Congress have expressed growing concern over the quality of grain exports. Foreign buyers have also complained. Grain trading such as we know it would probably not be possible without some sort of grading system. The 1985 Food Security Act commissioned the Office of Technology Assessment to study the quality issue. It also instructs the Federal Grain Inspection Service to develop new means of establishing grain classifications, taking into account characteristics other than those visually evident.

*** 4.30 ***

Hanrahan, Charles E., and T. Kelley White. *Consortium on Trade Research: Imperfect Competition, Market Behavior, and Agricultural Trade Policy Analysis*. Washington, DC: Economic Research Service, U.S. Department of Agriculture, 1983.

International markets for agricultural commodities are not perfectly competitive, and the usual assumptions underlying the perfectly competitive model of market behavior do not hold for these markets. The analysis suggests that government policymakers place different values on the welfare of various interest groups. Assumption of maximization of net social payoff in models of market behavior is inappropriate.

*** 4.31 ***

Johnson, D. Gale, Kenzo Hemmi, and Pierre Lardinois. *Agricultural Policy and Trade*. New York: New York University Press, 1985. ISBN 081471673, 081474168 (paperback).

The agricultural policies of the major industrial nations have deep internal roots. Price supports and liberal trade are not in conflict when support levels are set below the trend of international market prices and are adjustable as markets change. The prospects for changing the current domestic agricultural programs to market-oriented policies will be greatly enhanced if the macroeconomic policies in Europe, Japan, and the United States provide for employment conditions that make it relatively easy for farm men and women to find non-farm jobs when farm employment provides inadequate incomes. The authors emphasize that farm people cannot be expected to accept more market-oriented policies unless there are non-farm job opportunities of an attractive nature available.

*** 4.32 ***

Moulton, Kirby, and C. Parr Rosson, III. "Trading for Prosperity in American Agriculture: A Political Pipedream or a Practical Plan?" *Policy Choices for a Changing Agriculture*. North Central Regional Extension Publication. Columbus, OH: 1987.

World agricultural trade will increase over the next two decades in response to larger populations and higher incomes. However, the U.S. trade share will depend on costs of production and marketing relative to competitors, the impact of government policies affecting trade, and changes in the dollar's value. That share will be higher if costs are low and not offset by trade barriers or currency fluctuations or if government price subsidies are high and not matched by competitors.

Growth in World Trade

*** 4.33 ***

Sorenson, Vernon L. "International Food Policy and the Future of the Farm and Food System." *The Farm and Food System in Transition*, no. 9. East Lansing: Michigan State University. Cooperative Extension Service, 1983.

"American agriculture has become an international business," Sorenson asserts. U.S. and world markets are becoming increasingly integrated in direct consumption imports as well as in exports and imports that affect food industries and farming. Integration means that the markets for U.S. farm products and inputs are subject to political and economic changes around the world over which we have no control.

Alternative Marketing Strategies

*** 4.34 ***

Angell, George. *Agricultural Options: Trading Puts and Calls in the New Grain and Livestock Futures Markets*. New York: American Management Association, 1986. ISBN 081445822X. LC 85-48215.

Angell's book can best be described as a guide to profitable trading in agricultural options markets. He overemphasizes opportunities for profits without mentioning the chances for losses. However, the author does provide an interesting discussion of the views and strategies of a practicing professional trader.

*** 4.35 ***

Antonovitz, Frances, and Ronald Raikes. "An Improved Marketing Strategy for Producers Who Use Direct Marketing." *North Central Journal of Agricultural Economics* 5:1 (January 1983):31-36.

Producers of fed cattle who use direct marketing habitually contact only two or three packers to obtain a bid when selling their cattle. The authors used search theory to determine an improved direct marketing strategy. Two models of this strategy were developed. Results from an empirical example show that depending on the producer's subjective distribution of packers' bid prices, expected returns may be higher if this improved strategy is used.

*** 4.36 ***

Carlton, Dennis W. "Futures Trading, Market Interrelationships, and Industry Structure." *American Journal of Agricultural Economics* 65:2 (May 1983): 380-387.

Futures markets arise as a response to economic uncertainty. Carlton attempts to measure how futures trading in existing contracts changes in response to changes in uncertainty caused by inflation. If there were no uncertainty, there would be no need for futures markets. Rather than diversifying risk through many activities, firms can focus attention on the activity they do best and use a futures market to diversify their risk.

*** 4.37 ***

Forker, Olan D., and V. James Rhodes, eds. *Marketing Alternatives for Agriculture, Is There a Better Way?* National Public Policy Education Committee. Publication, no. 7. Ithaca: New York State College of Agriculture and Life Sciences, Cornell University, 1976.

Rhodes, V. James, and Olan D. Forker. *The Situation Now*, 7-1.

Henderson, Dennis R., Lee F. Schrader, and Michael S. Turner. *Electronic Commodity Markets*, 7-2.

Sporleder, Thomas L., and David L. Holder. *Vertical Coordination through Forward Contracting*, 7-3.

Holder, David L., and Thomas L. Sporleder. *Forward Deliverable Contract Markets*, 7-4.

Moulton, Kirby, and Daniel I. Padberg. *Mandatory Public Reporting of Market Information*, 7-5.

Shaffer, James D., and Randall E. Torgerson. *Exclusive Agency Bargaining*, 7-6.

Black, William E., and James E. Haskell. *Vertical Integration through Ownership*, 7-7.

Myers, Lester H., Michael J. Phillips, and Ray A. Goldberg. *Joint Ventures between Agricultural Cooperatives and Agribusiness-Marketing Firms*, 7-8.

Armbruster, Walter J., Truman F. Graf, and Alden C. Manchester. *Marketing Orders*, 7-9.

Abel, Martin E., and Michele M. Veeman. *Marketing Boards*, 7-10.

Garoyan, Leon, and H. M. Harris, Jr. *Industrial Restructuring: A Policy for Industrial Competition*, 7-11.

Knutson, Ronald D., Dale C. Dahl, and Jack Armstrong. *Fine Tuning the Present System*, 7-12.

Knutson, Ronald D., and Olan D. Forker. *The Options in Perspective*, 7-13.

"It would be easy to overstate the extent of new agricultural marketing problems. Some are as old as the country itself," Rhodes and Forker point out in the introduction to this series of leaflets. In 1976 a group of agricultural economists raised the question about marketing farm commodities under the theme "Is there a better way?" The search for a better way goes down many avenues as important and sometimes impossible demands are made on the performance of the marketing system. New marketing institutions have developed from time to time to meet perceived problems. Yet institutions also flourish and phase out over time. The sophisticated systems of agribusiness to procure farm products include a growing degree of vertical coordination and raise policy and marketing issues discussed in this series.

*** 4.38 ***

Futtrell, Gene A., ed. *Marketing for Farmers*. St. Louis: Doane Western, 1982. ISBN 0932250181. LC 82-70257.

Terms like "marketing strategy" and "risk management" were heard more frequently after the early 1970s when agriculture found itself in a new era of market volatility. This book is a marketing guide for farmers to help them make those decisions so crucial to continued livelihood. Topics include determining grain marketing objectives, grain storage as a marketing strategy, selling through cash contracts, futures markets, sources of information, developing a livestock marketing plan, using livestock futures, sources of information on livestock markets, and technical price analysis.

*** 4.39 ***

Henderson, Dennis R. "Electronic Marketing in Principle and Practice." *American Journal of Agricultural Economics* 66:5 (December 1984):848-853.

The basic concept of electronic marketing is simultaneous trade negotiations among separated buyers and sellers channeled into an interactive central market through electronic communications. Product movement occurs later. Neither traders nor products are physically assembled at a common location; products are sold by description rather than personal inspection by the buyer. Electronic marketing can be viewed as offering the potential to mitigate perceived pricing problems and improve coordination in agriculture without imposing higher exchange costs on participants. So far experience shows a lack of resounding commercial success.

*** 4.40 ***

Kinnucan, Henry, and Olan D. Forker. "Seasonality in the Consumer Response to Milk Advertising with

Implications for Milk Promotion Policy." *American Journal of Agricultural Economics* 68:3 (August 1986):562-571.

Milk advertising expenditures evenly distributed over the year, with only small fluctuation occurring in concert with demand shifts, gain corroborative support at both the theoretical and empirical level. The authors conclude that appropriate timing of milk advertising expenditures can increase the effectiveness of the investment. Monthly changes in the farm value of Class I milk, the Class I utilization rate, and the ability of milk advertising to influence sales affect the profitability of the advertising investment.

*** 4.41 ***

Nelson, Ray D. "Forward and Futures Contracts as Pre-Harvest Commodity Marketing Instruments." *American Journal of Agricultural Economics* 67:1 (February 1985):15-23.

Empirically significant differences between forward and futures contracts illustrate their imperfect substitutability as pre-harvest marketing instruments. Certain combinations of market conditions make the two types of contracts complementary rather than interchangeable. Basis constitutes the most important difference between forward and futures prices. Practically speaking, large traders find only basis as a significant difference between forwards and futures when comparing the two contracts as marketing alternatives.

*** 4.42 ***

Riggins, Steven K., Jerry R. Skees, and Michael R. Reed. "Analysis of Grain Marketing Pricing Strategies." *Journal of the American Society of Farm Managers and Rural Appraisers* 50:1 (April 1986):58-64.

The increased reliance on international markets has made U.S. grain prices more volatile. This has put additional pressure on farmers to improve their marketing skills. Many producers have marketing plans based on tradition and preconceived ideas. Frequently these plans or traditions result in consistently low net prices. Marketing economists have argued that producers who utilize current marketing information will receive a higher net price and possibly reduce the risk in marketing relative to producers who don't use this information.

*** 4.43 ***

Roy, Ewell Paul. *Collective Bargaining in Agriculture*. Danville, IL: The Interstate Printers & Publishers, 1970. LC 79-113823.

Roy explores the ways and means of increasing farmers' bargaining power in the marketplace. Despite all the research and educational and technological advancement in agriculture, the lack of bargaining strength held by farmers remains a major unsolved problem. The farmer's need for more bargaining power is recognized, but reaching agreement on how to attain it remains the key problem. Although farm organizations agree that farmers should organize bargaining associations to obtain better prices, they are divided on how to achieve it as they are on other issues. The shift in political power away from farmers along with a rising coalition of food processors, wholesalers, and retailers will make legislation for farm bargaining more difficult.

*** 4.44 ***

Russell, James R., and Wayne D. Purcell. "Participant Evaluation of Computerized Auctions for Slaughter Livestock--the Experience with Electronic Marketing Association, Inc." *North Central Journal of Agricultural Economics* 6:1 (January 1984):8-16.

Attitudes and perceptions of buyers and sellers, who have used Electronic Marketing Association's computerized trade system, were examined. Participants in both the unsuccessful Virginia slaughter cattle program and the successful Eastern Lamb Producer Cooperative lamb program were interviewed. Both sets of respondents generally indicated a positive attitude toward computerized trading. Inferences were made about causal factors affecting successful implementation of computerized trading systems. Applicability of the findings of this study across commodities systems or participants will require discretion, but the general directives concerning procedure should be broadly applicable.

*** 4.45 ***

Torgerson, Randall E. *Producer Power at the Bargaining Table*. Columbia: University of Missouri Press, 1970. LC 76-113817.

"Farm operators are playing in a new political ball game," Torgerson concludes as he assesses the significance of S. 109, the major focus of this book on farmer bargaining. Farm bargaining was inspired by the success of organized labor. As unions acquired strength, unorganized farm producer groups felt they could gain by organized bargaining for prices of their products. This study tells the story of Senate Bill S. 109, an effort to authorize farm bargaining, and the issues associated with farm marketing in the late 1960s. In the process, the author also illustrates the changing balance of farm power on Capitol Hill. He concludes that the greatest challenge in achieving further bargaining legislation is for farm groups to unite with each other and with nonfarm groups to attain desired objectives. He believes that farm operators can still "pack a wallop" when their organizations are united.

*** 4.46 ***
Ziemer, Rod F., and Fred C. White. "Disequilibrium Market Analysis: An Application to the U.S. Fed Beef Sector." *American Journal of Agricultural Economics* 64:1 (February 1982):56-62.

Ziemer and White consider the nature of price adjustments in agricultural commodity markets in light of recent developments in disequilibrium theory and estimation. By disequilibrium they mean that market transactions occur at prices that do not clear the market, so that some buyers or sellers are not able to trade desired quantities at the prevailing price. In general, there are numerous possible causes of disequilibrium such as inordinate weather, sudden changes in population, or other demographic variables and specific influences peculiar to a particular market. Although this type of analysis is in an early stage of development, specification and estimation of relatively simply models are possible. The model developed by the authors yielded more accurate fed beef price forecasts than the equilibrium model.

Policies Related to Trade

*** 4.47 ***
Ballenger, Nicole, John Dunmore, and Thomas Lederer. *Trade Liberalization in World Farm Markets*. Agricultural Information Bulletin, no. 516. Washington, DC: Economic Research Service, U.S. Department of Agriculture, 1987.

"The playing field in world agricultural trade is uneven," the authors assert. Some countries subsidize their imports; others subsidize their exports. Most nations try to help their farmers through domestic farm programs. All these influences converge to distort the price signals that would otherwise govern supply and demand in a freer trade environment. The resulting hodgepodge has disadvantaged many producers in the United States and other countries, even though their farm products are among the least expensive to produce. This report gives a perspective on the issues and conditions that the United States and the other 91 members of the General Agreement on Tariffs and Trade confront as they negotiate reduction in intervention in international markets.

*** 4.48 ***
Blandford, David, and Richard N. Boisvert. "Employment Implications of Exporting Processed U.S. Agricultural Products." *American Journal of Agricultural Economics* 64:2 (May 1982):347-354.

Less attention has been given to the question of whether the United States is realizing the maximum gains for the economy as a whole from foreign market opportunities. In assessing the economy-wide benefits of export expansion, one would need to estimate the net returns to all factors of production attributable directly or indirectly to agricultural sales overseas. The number of jobs generated depends not only on total volume of agricultural exports, but also on the degree to which they are processed prior to export. It is particularly important to identify the effectiveness of promotional expenditures in expanding the demand for both primary and processed products.

*** 4.49 ***
Brada, Josef C. "The Soviet-American Grain Agreement and the National Interest." *American Journal of Agricultural Economics* 65:4 (November 1983):651-656.

In the 1970s, the Soviet Union became a major importer of grain, giving rise to fears that it could extract an undue share of the gains from East-West trade through the monopolistic power of its state-trading organs and by keeping its buying intentions secret. Offer curves depicting Soviet trade are constructed. The author shows that the grain agreement between the two countries negates most of the bargaining advantage attributed to the Soviet trade monopoly and may open the Soviet Union to exploitation by the United States.

*** 4.50 ***
Gilmore, Richard. *A Poor Harvest: The Clash of Policies and Interests in the Grain Trade*. New York: Longman, 1982. ISBN 0582281938. LC 81-19309.

Although most of the sound and fury surrounding the international grain trade has died down, serious questions about international trade and domestic agricultural policies remain. In this book, Gilmore attempts to analyze the interactions between the international grain trade and domestic agricultural policies. He concludes "what now exists is a world in which economic and political power rather than market forces are the prime determinants which match available food supplies to demand."

*** 4.51 ***
Krueger, Anne O. "Protectionism, Exchange Rate Distortions, and Agricultural Trading Patterns." *American Journal of Agricultural Economics* 65:5 (December 1983):864-871.

The world has focussed on liberalizing trade in manufactures and failed to react to apparently increasing distortions in trade in agricultural commodities. The problems confronting liberalization of trade in agricultural and nonagricultural commodities have become increasingly similar over the past several decades. The challenge to economists and policy makers will be to devise means of achieving trade liberalization from non-tariff barriers.

*** 4.52 ***

National Commission on Agricultural Trade and Export Policy. *New Realities: Toward a Program of Effective Competition.* Final Report to the President and Congress, vols. 1 & 2. Washington, DC: U.S. Government Printing Office, 1986.

"The continuing decline in U.S. agricultural exports is sending a danger signal to the U.S. economy," this report warns in the opening summary. U.S. agricultural exports and trade are vital factors in the well-being of our nation's economic system. Employment for one-fifth of our nation's work force is linked to our ability to maximize opportunities arising in the domestic and world market for basic agricultural commodities and value-added products. A variety of factors has contributed to the recent decline in U.S. agricultural competitiveness. Competition in world markets results from the direct actions of government. U.S. domestic policies, both agricultural and nonagricultural, have had an important inhibitory effect on our nation's competitive position.

*** 4.53 ***

Purcell, Randall B., and Elizabeth Morrison, eds. *U.S. Agriculture & Third World Development, the Critical Linkage.* A Curry Foundation Policy Study. Boulder, CO: Lynne Rienner Publishers, 1987. ISBN 1-55587-011-2.

"Agricultural trade policy is buffeted by other, larger forces that move the modern economy. Exchange rates, interest rates, and debt payments have more to do with the volume and value of U.S. agricultural trade than does domestic agricultural or agricultural trade policy," Rossmiller and Tutwiler conclude as they look at agricultural development and trade. The trend toward insularity in international affairs and protectionism in international trade are public concerns that the sponsors of this volume wanted to address. The writers try to demonstrate that the United States can export to restore full health to its agricultural sector only as developing countries improve their own agriculture. They call for those responsible for influencing U.S. farm and agricultural trade legislation and U.S. policy toward developing countries to work together to achieve mutual goals.

*** 4.54 ***

Smith, Mark. *Increased Role for U.S. Farm Export Programs.* Agricultural Information Bulletin, no. 515. Washington, DC: Economic Research Service, U.S. Department of Agriculture, 1987.

Lower prices, commercial credit, and food aid programs are important tools for assisting U.S. exports of a variety of agricultural commodities. Export programs of the Commodity Credit Corporation that use those tools have helped maintain or expand U.S. market shares in some countries, but they may not be sufficient to offset other factors that may weaken U.S. market position. The drop in U.S. agricultural exports in the early 1980s raises the question: how important and effective are U.S. programs in boosting agricultural exports?

*** 4.55 ***

Sorenson, Vernon L., and George E. Rossmiller. "Future Options for U.S. Agricultural Trade Policy." *American Journal of Agricultural Economics* 65:5 (December 1983):893-900.

During the 1970s unprecedented growth occurred in the international market for agricultural products. This growth had a profound impact throughout the U.S. food system and resulted in a phenomenal increase in interdependence within and among trading nations. Agricultural trade policy is almost always forced into the role of supporting domestic policy due to the assumed sovereign right of most governments to follow farm policies that reflect primarily their domestic interests. The new need in trade policy is to recognize the nature of international and intersectoral interdependence and to develop a set of mechanisms to seek collaborative approaches to problems of international market performance. The search is needed for a more collaborative alternative that seeks to deal with the mix of domestic and international policy in the perspective of mutual national interests. The globally interdependent food system is a product of the post World War II period. A global food system has existed only since the recent development of a world linked by sophisticated communications systems, international markets, rapid and relatively inexpensive intercontinental transportation systems, and a scientifically-based producing system.

*** 4.56 ***

United States. Economic Research Service. *Government Intervention in Agriculture, Measurement, Evaluation, and Implications for Trade Negotiations.* FAER-299. Washington, DC: U.S. Department of Agriculture, 1987.

"Without exception and with varying degrees of comprehensiveness and success, all governments intervene in agriculture," the authors declare. This study analyzed government intervention in the agricultural sectors of the market-oriented countries most active in trade. Levels of assistance or taxation to agricultural producers and to consumers in the form of domestic farm programs and agricultural trade barriers are measured by producer and consumer subsidy equivalents for 1982-84.

*** 4.57 ***

United States. Department of Agriculture. *U.S. Agriculture in a Global Economy.* 1985 Yearbook of Agriculture. Washington, DC: U.S. Government Printing Office. LC 85-600627.

The world is shrinking--politically, if not physically. The policies adopted by one country may affect the affairs of many. This yearbook provides a comprehensive picture of agriculture in an interdependent world and the relationships between domestic policies and international trade practices. Agricultural technological progress will continue, and the rate may actually increase over the next two decades.

*** 4.58 ***

Wisner, Robert N., and Tahereh Nourbakhsh. *World Food Trade and U.S. Agriculture, 1960-85.* 6th ed. Ames: World Food Institute, Iowa State University, 1986.

World grain and oilseed production set new records in 1985-86, while grain exports fell to the lowest level since 1978-79. That produced rapidly rising world carryover stocks; sharply increased cost of government farm programs in the European Community, United States, and other countries; a continued drift toward trade protectionism and trade policy conflicts; and one of the most dramatic changes in U.S. farm policies in half a century.

Marketing and Public Policy Issues

*** 4.59 ***

Armbruster, Walter J., and Lester H. Myers, eds. *Research on Effectiveness of Agricultural Commodity Promotion.* Oak Brook, IL: Farm Foundation, 1985.

Up to this time agricultural economists had paid little attention to how effective agricultural commodity promotion programs were. This book examines information needs; reviews completed research; examines conceptual, empirical, analytical, and measurement issues; suggests policy implications and research directions; and provides an overview of where the action is in evaluation of agricultural commodity promotion programs. The action is likely to get more intense as surpluses grow, prices are depressed, and producers are forced to grapple with ways to expand markets. Promotion efforts represent an approach that will receive increasing attention.

Preparing for the Future

*** 4.60 ***

Farris, Paul L., ed. *Future Frontiers in Agricultural Marketing Research.* Ames: Iowa State University Press, 1983. ISBN 0913805600. LC 82-70257.

The editor provides a comprehensive survey of emerging and persisting research problems in the field of agricultural marketing. The book is designed to develop perspectives for research priorities in marketing, to conceptualize marketing problems in the contemporary economic setting, to suggest research approaches in various situations, and to outline the resources necessary to solve emerging problems. The authors of the 16 papers include most of the perceived heavyweights in agricultural marketing.

General References

*** 4.61 ***

Branson, Robert E., and Douglass G. Norvell. *Introduction to Agricultural Marketing.* New York: McGraw-Hill Book Co., 1983. ISBN 0070072418. LC 82-18016.

In this book intended for undergraduate agricultural marketing courses, the authors combine a point of view traditionally taken by agricultural economists with the view more common in business schools. They include an industry-wide analysis of the marketing process and a firm-level treatment heavily influenced by economic theory. Extensive discussions include market structure, price theory, contract markets, government involvement in marketing, farmer, wholesaler, and retailer marketing strategy, product development, packaging, promotion, international sales, and farm input marketing.

*** 4.62 ***

Kohls, Richard L. *Marketing of Agricultural Products.* New York: Macmillan Company, 1955. LC 55-717.

In this agricultural marketing textbook, Kohls provides a basic description of marketing. He looks at marketing partly by function, partly by institution, and partly by commodities. He points out that those who work in marketing are contributing to form and place utility and are not parasitic to the system as some who may decry the "profits of middlemen."

*** 4.63 ***

Kohls, Richard L., and Joseph N. Uhl. *Marketing of Agricultural Products.* 6th ed. New York: Macmillan Publishing Company, 1985. ISBN 0023656700. LC 84-817.

Comprehensive coverage makes this revised textbook useful reading for anyone interested in the organization and functioning of the food marketing system. The authors present some controversial topics such as advertising, market power, marketing costs, and market regulations in a balanced manner. The interaction and potential conflict among performance measures such as operation efficiency, pricing efficiency, competition, and consumer satisfaction are woven through most of the twenty-nine chapters.

*** 4.64 ***

Rhodes, V. James. *The Agricultural Marketing System.* 3d ed. New York: John Wiley & Sons, 1987. ISBN 0471851000.

In this agricultural marketing text in three parts, Rhodes addresses product and production characteristics, marketing-agribusiness functions, domestic consumption and demand, international markets, transportation and storage, the futures market, pricing and exchange, grading, co-ops, market orders, and the interfaces between processors, wholesalers, retailers, and consumers. In content and format, the book is a major revision of the first edition, which appeared in 1978.

*** 4.65 ***

United States. Department of Agriculture. *Food -- from Farm to Table.* 1982 Yearbook of Agriculture. Washington, DC: U.S. Government Printing Office, 1982. LC 82-600649.

Farmers and supporting activities and industries make up an agricultural system intertwined with the general economy. The system generates about 20 percent of the Nation's gross national product, and employs 23 percent of the U.S. labor force. Yet only 3 percent of the labor force is engaged directly in farming. The 1982 Yearbook provides a panorama of the changing economics of agriculture, farm marketing in a new environment, and food buying.

*** 4.66 ***

United States. Department of Agriculture. *Contours of Change.* The 1970 Yearbook of Agriculture. Washington, DC: U.S. Government Printing Office, 1970.

In this volume, which looks at many phases of the agricultural revolution, several chapters provide useful explanations of the economics of agricultural marketing. Topics of special interest include the marketing system, marketing orders, marketing costs, foreign trade, and world trends in use of farm products. The continued growth of population and metropolitan areas is viewed as an environment of opportunity for expansion in economic activity in rural areas.

CHAPTER 5

THE FARM AND ITS SETTING IN THE RURAL COMMUNITY

"Changes in technology tend to precede changes in social and cultural ways of life."
William Ogburn[1]

The connection between the American farm and the surrounding community has changed profoundly. The farm input market has become an important part of the business of the farming community. Many rural communities have also developed non-farming industries and many farmers and their families benefit from employment off the farm.

The connection between the American farm and the community has changed profoundly over the years. Today, more than ever, farms are not completely isolated units. Rather they are located adjacent to other farms, along roads, and are a part of the rural community setting. No longer are individual farms completely self-sufficient.

The purpose of this chapter is to show the influence of farming upon the surrounding local community, how non-farming businesses and industry affect farms, and how changes in the nature and structure of farms and the farm financial problems experienced by farmers in the 1980s affect the rural community.

The Rural Community Setting

Two basic trends have clearly been operating in rural areas for several decades. First, labor has been displaced from agriculture primarily due to mechanization and consolidation of agricultural production. Second, the number of manufacturing and service sector jobs has grown in rural areas.

The industrialization of rural America increased greatly in the 1970s, especially in the South. Reduction in the population of farm operators slowed in the 1970s and reversed in some areas. Overall, rural areas in the United States are much less dependent on farming in the 1980s than in the 1940s. In 1983, the natural resource sector, which includes all farming operations and related businesses, as well as forestry, fisheries, and mining, accounted for only 11 percent of the wage and salary income of nonmetropolitan areas.

Also the gap between incomes of rural and urban workers narrowed in the 1970s. However, the poorest counties in the United States are nearly all nonmetropolitan. Some regions of the country have improved rural welfare more than others.

The changes in the American farm from its traditional small-scale mode to a modern industrialized form affect the structure of our farms and all of the surrounding rural community.

The Changing Rural Community

By definition, everything that is not defined as urban is considered rural. Cities of 2,500 or more population are considered to be urban. Cities with a population of 50,000 or more along with the surrounding area are defined as standard metropolitan statistical areas.

Up until World War II, farming and farm related industry was usually the dominant economic activity in rural America. But 30 years of farm consolidation, widespread outmigration of farm and non-farm people from rural areas dependent on farming, and growth of rural non-farm jobs have changed the rural community.

In the late 1980s, most rural people do not depend on farming as a principal source of employment or income. Even farmers are less dependent on agriculture: around 60 percent of total farm family income now comes from non-farm sources. In 1980, about one-fourth of the U.S. population, 59.4 million persons, lived in rural areas, but the majority of these people were employed in occupations not directly tied to farming. Although 5.6 million people lived on farms, they comprised only 2.5 percent of the total U.S. population.

Only about 10 percent of labor and proprietors' income in nonmetropolitan areas is attributable directly to farming. In 702 nonmetro counties out of a total of 2,443, a significantly greater direct dependence on farming has been observed. In about a third of these farming dependent counties, farming accounted for nearly 50 percent of farm labor and proprietors' income.

Farming, Employment, and Related Businesses

Farming affects the employment and related businesses in the rural community. Farming, as well as farm input industries, farm processing industries, and food and fiber wholesaling and retailing all provide employment. In 1982, the food and fiber system accounted for more than 30 percent of the jobs in nonmetro areas.

Because today's farms are increasingly dependent on purchased inputs, the farm input market is an important form of interaction between farming and the rural community economy. A substantial portion of the increase in farm productivity is due to increased use of physical capital, energy, fertilizers, chemicals, improved seed varieties, and animal breeding services. The threat of shortages of fuel and nitrogen fertilizers in the mid-1970s illustrate the linkage and influence of the non-farm economy upon farm production practices and management decisions.

However, the farming economy no longer dominates rural or nonmetropolitan America. Rural communities are rapidly supplementing farming with non-farm activities. The diversity among rural counties has increased since World War II as farm numbers declined, farmers and farm families moved into other employment, and new business and industry developed. The reasons for a resurgence in many rural areas are not the same in all areas. One reason is the slowing of structural changes in agriculture. The decline in farm numbers has slowed. Further outmigration from farms has been reduced or eliminated.

In the more diversified rural economies, the availability of non-farm jobs and employment creates opportunities for farm families and permits many to continue farming on operating units that would be too small to support full-time employment on the farm alone.

New economic activities have moved into rural areas and diversified the sources of employment. Trade, services, and manufacturing are far more important than before.

Rural isolation has decreased and many people now view the remaining isolation from cities and cultural differences as advantages rather than disadvantages. Improved communication, transportation, and employment opportunities now enable many non-farm citizens, who might otherwise live in metropolitan suburbs, to live comfortably in rural areas.

Industrial diversification into rural areas has provided farmers and farm families the opportunities for employment and survival as part-time farmers. Dependence on off-farm income has increased rapidly, rising from 42 percent of total farm family income in 1960 to more than 60 percent in 1985. Farm families with annual farm sales of $40,000 to $100,000 earn over 60 percent of their total yearly income from off-farm sources. Farm families with less than $40,000 in annual farm sales derive essentially all of their net income from sources other than farm production. Many farmers with sales of less than $20,000 annually are deliberately and permanently part-time farmers and are virtually full-time workers somewhere in the rural economy.

A new environment has emerged in the relationships between farming and the rural community. Part-time farming is no longer a transitional stage during which farmers and their family members take off-farm work on their way into or out of full-time farming. Instead, such part-time farming has come to represent a permanent and important part of a stable multi-job rural career.

Farmers with less than $40,000 in annual sales comprise just over 70 percent of all farms. Many of these operators have good incomes from off-farm sources. But many farms with annual sales between $40,000 and $100,000 face possible extinction because the families are more limited in the time or opportunities they have to develop off-farm income sources. Farms will be larger or smaller, as the demand of daily farm operations and the need for substantial outside income on farms of this size are increasingly incompatible.[2]

The Influence of Changing Agriculture on the Rural Community

Agricultural structure is related to different types of social interactions in rural communities. With the decline in the number of farms in the United States since 1935, a dual agricultural system has begun to emerge with a few large farms and a large number of small farms.

An early study in California in the 1940s suggested that the community with large farms and a hired labor force would have a lower quality of life than a community surrounded by family farms as measured by family income, social and religious institutions, and the degree of local control of government services.[3] A broader study in the San Joaquin Valley of California in the 1970s concluded that the smaller scale farming areas tended to offer more to local communities than their larger counterparts. But these results do not necessarily apply to many other areas.[4]

Studies in the Midwest and Southeast have focused on differences between farms producing poultry and eggs under contract and those farms with independent operators and those involved in contract farming. These studies concluded that there was little evidence that the types of poultry producers were associated with different community status levels.[5]

Another study showed little difference in the community activities of those working in contract farming and family farming systems.[6] But there was a significant difference in the social participation of the workers and managers in larger-than-family farms. The managers were more involved in community and political activities than were the workers. A follow-up study between those producing poultry under contract and those producing feeder calves for a relatively competitive market showed that beef producers were slightly more involved in social activities in the community although poultry producers received slightly higher income from farming.[7]

A Maryland study revealed that as the scale of dairy farm operation increases, changes in the operational structure and functioning of the farm take place, which have negative effects on the ability of the farm family to participate in the formal social organizations of the community.[8]

Trade Patterns Related to Farming

A second community issue raised by changes in the size and structure of agriculture focuses on trade patterns. The question raised is whether large scale farm operators operate differently in their purchases of farm supplies and their sale of commodities than small scale operators. An unpublished 1969 Louisiana study showed few differences in the distance that family farmers, contract farmers, or larger than family farm owner-managers and their families traveled for the purchase of farm or personal goods and services. Likewise there were not many differences in 1981.[9]

A study in three Iowa areas concluded that there were very few significant differences in the size of acres operated and gross farm income between farmers who purchased in small local trade centers and farmers who purchased in large distance trade centers.[10] Where differences existed, large scale farmers were as likely to purchase in small, local trade centers as were small-scale farmers.

Although the number of farm suppliers and farm commodity markets are expected to decline, the researchers believed that no trend toward "agri-megacenters" that would concentrate agricultural supply centers in a few locations was developing in the 1980s.

However, the increasing need for specialized knowledge, skills, and equipment to service farm production will prevent every small equipment, chemical, and marketing facility from providing a full array of services. The specialized services needed by industrialized farmers will be provided by a few larger, centralized facilities and may be directly tied to satellite facilities in other rural communities. The result may be that control and some of the income from businesses and services for farmers may be drained from many smaller rural communities.

Competition and Conflicts over Land Use

Since World War II, farm land has become the focus for several major changes in American society. Most of these changes became more noticeable during the 1970s. The major issues centered on urban demand for farm land for development use, recreational demand for rural land, and government programs that required land acquisition for such projects as the interstate highway system, water development projects, and airports.

At the same time, especially in the 1970s, inflation and fears of world food shortages were causing a strong demand for farm land as an investment. Many were concerned about the loss of farm land to non-farm uses. The increased migration into nonmetropolitan areas also brought demands for land to build homes and establish rural residences.

The increase in rural non-farm population among traditional farming communities brought conflicts between farmers and nonfarmers over environmental questions: noise, appearance, and odors that accompany crop and livestock production. Out of these concerns emerged various efforts to protect farm operators and farmland owners: rural zoning to maintain farms where urban pressures might otherwise force a shift of the land into non-farm uses, establishment of development rights that would preserve certain open spaces for farming, and use-value assessment procedures to prevent rising real estate property tax rates from forcing farmers from the land.

As cities expand, the suburban-front surrounding a city spills out into open farmland affecting land values, breaking up established farming units as land is sold for development, and raising land values to the point where it can no longer be purchased for farming uses. As the acreage in farms declines in the growing suburban community, the businesses that have traditionally served the farming community must redirect their business services to non-farm residents or leave for lack of farmer customers.

In the 1970s the federal government expanded its policy to protect the environment. The American farm is both a source and a "victim" of environmental degradation. Farm production affects the environment in a variety of ways. Nationally, nonpoint water pollution, which originates from farm sources, causes conflict between farmers and the local community. Air and water pollution in rural areas, as well as climatic changes caused by air pollution affect farm productivity. Future policies of federal and state governments can be expected to affect certain farming practices and the methods that farmers must follow to reduce air and water pollution.

Closely related to environmental concerns are soil and water conservation efforts. Our national policy to conserve soil and prevent soil erosion dates back to the 1930s. The Soil Conservation Service was established by Congress in 1935. Since the 1970s, additional emphasis has been placed on water conservation and prevention of water pollution in rivers and streams. Nonpoint pollution, the runoff from fields, has become a special problem with increased use of fertilizers and chemicals in farming operations.

Farming and the Demand for Public Services

Services provided to farms and farm families are part of the state, county, and local government services to the entire community; schools, roads, highways, fire and police protection, park and forest preserve districts, and libraries.

As farmers today share their community increasingly with nonfarmers, the demands for public services may increase, placing pressure on local government units to expand these services. When the major share of local government revenues come from property taxes, conflicts have tended to arise among farmers, farmland owners, and the non-farm community.

The state and local public finance systems in the farm-dominated states are generally more decentralized and thus relatively more dependent on the property tax. Local governments serving farmers are usually smaller since they service fewer people. Local government units serving rural communities face some common problems.

Decreases in federal revenue sharing funds affect small local governments that may find it difficult to replace these funds with local sources.

Intergovernmental obligations placed upon local governments through mandates by federal and state governments place further pressures on local governments to raise revenues or reduce services.

New technology in agricultural production causes increased needs for certain public services. Roads and bridges, built in the early 1900s when agricultural transportation was limited to the horse and wagon, are inadequate for the 1980s and 1990s. Technological advances have brought increased numbers of delivery trucks for feed, fertilizer, petroleum and other inputs to every farm. Bulk milk trucks and other specialty trucks have become larger.

Financing of public schools is a major problem facing many rural communities where populations are small and school costs per pupil are higher. In farm states with declining land values, the ability to finance improved public schools is limited since the property tax has been the major source of these revenues.

Financial Stress in Farming and the Rural Community

Close observers of the rural community see the farm crisis of the 1980s as only one of the most visible manifestations of a fundamental transition occurring in American society as we shift from a mass society to an information age. Accompanying this change will be changing attitudes about government help for farmers and difficult decisions about how to provide institutional services in those local government units dependent upon real estate property tax revenues that are shrinking with reduced farm incomes and reduced property assessments.

The financial stress experienced by the farm economy is expected to affect the performance of state and local governments. Depending on the importance of farming to the state and local economies and the structure of governmental finance, the economic stress on farms will adversely affect the performance of the tax systems and the provision of a broad range of local government services.

Decreases in property tax assessments will reduce the capacity of rural property tax bases and increase the stress on those rural local governments most dependent upon property taxes for their revenue. School districts, townships, and county governments may be most directly affected. Communities where farming is an important basic economic activity and a major component of tax bases are likely to be most seriously affected.

As middle-sized farmers disappear, many are concerned that the businesses along main street that have served the segment of the farming community will also go with them.

Conclusions

The American farm is an economic unit in an industrialized society where raw materials are brought to the farm production plant, combined with land to produce crops, or used for animals to produce meat, dairy, and poultry products. When ready for sale, the commodities are shipped off to a market for further processing into food and fiber products.

Farms cannot operate in isolation. They provide employment to local suppliers and service industries and raw materials for processing and manufacture.

Many farming operations are conducted in nonmetropolitan counties that also have industrial plants and non-farm industries operating adjacent to farms and farming communities. The mix of farm and non-farm activity opens the way for conflict and the need for cooperative efforts to maintain compatible relationships. With the decline in farm numbers and reductions in the number of farm service firms, more and more effort will be made to attract other industries into rural communities.

Farms and farm families require public services just as non-farm residents. And with increasing numbers of nonfarmers in the rural communities, farmers, farmland owners, and non-farm residents must share the cost of public services required.

Changes in agricultural technology will continue to change the demands for farm inputs, public services, numbers of farm operating units in the rural community, and participation by farm operators and farm workers in the social organizations of the rural community.

The financial problems faced by many farm families will force changes in the number of farm supply and service businesses and place strains on state and local governments to provide public services in states and communities where farming is important. How to deal with this and other problems will be the focus of our three concluding chapters.

Endnotes

1. Wayne D. Rasmussen. "Agricultural and Rural Policy: A Historical Note." In *New Dimensions in Rural Policy: Building upon Our Heritage.* Joint Economic Committee, U.S. Congress, S. Prt. 99-153. Washington, DC: U.S. Government Printing Office, 1986. p.45.

2. Richard J. Sauer. "Agriculture and the Rural Community: Opportunities and Challenges for Rural Development." In *Interdependencies of Agriculture and Rural Communities in the Twenty-first Century: The North Central Region, Conference Proceedings.* Ames, IA: North Central Regional Center for Rural Development, 1986.

3. William D. Heffernan and Rex R. Campbell. "Agriculture and the Community: The Sociological Perspective." In *Interdependencies of Agriculture and Rural Communities in the Twenty-first Century: The North Central Region.* Ames, IA: North Central Regional Center for Rural Development, 1986.

4. Robert L. Moxley. "Agriculture, Communities, and Urban Areas." In *New Dimensions in Rural Policy.* Joint Economic Committee, U.S. Congress. Washington, DC: U.S. Government Printing Office, 1986. p.322.

5. Moxley. op. cit. p.323-325.

6. Heffernan and Campbell. op. cit. p.42.

7. Heffernan and Campbell. op. cit. p.42.

8. Moxley. op. cit. p.323.

9. Heffernan and Campbell. op. cit. p.43.

10. Ron Shaffer, Priscilla Salant, and William Saupe. "Understanding the Synergistic Link between Rural Communities and Farming." In *New Dimensions in Rural Policy: Building upon Our Heritage.* Joint Economic Committee, U.S. Congress. Washington, DC: U.S. Government Printing Office, 1986. pp.308-322.

Chapter 5: Annotated Bibliography

The Rural Community Setting

*** 5.1 ***

Bender, Lloyd D., Bernal L. Green, Thomas F. Hady, John A. Kuehn, Marlys K. Nelson, Leon B. Perkinson, and Peggy J. Ross. *The Diverse Social and Economic Structure of Nonmetropolitan America.* Rural Development Research Report, no. 49. Washington, DC: Economic Research Service, U.S. Department of Agriculture, 1985.

The authors identify seven distinct types of rural counties according to their major economic base, presence of federally-owned land, or population characteristics. The 702 counties identified as primarily dependent on farming are extremely rural, have stable populations with high per capita incomes and a highly capitalized farming industry, and are sensitive to agricultural policies, changing interest rates, and foreign trade. Other groups are those primarily depending on manufacturing, mining, and government functions; those largely occupied by federal lands or retirement settlements; or those characterized by persistent poverty.

*** 5.2 ***

Paarlberg, Don. "Rural America in Transition." In *New Dimensions In Rural Policy: Building upon Our Heritage*. S. Prt. 99-153. Joint Economic Committee, Congress of the United States. Washington, DC: U.S Government Printing Office, 1986.

Agriculture is being changed from its traditional small-scale mode to a modern industrialized form," Paarlberg observes. In this perspective paper, he points out that these changes are having a far-reaching effect on the structure of our farms and on all of rural America. Agriculture, formerly distinct from the non-farm sector in both culture and structure, is losing its uniqueness and entering the mainstream of economic, social, and political life.

*** 5.3 ***

Petrulis, Mindy F. "Effect of U.S. Farm Policy on Rural America." *Rural Development Perspectives* 1:3 (June 1985):31-33.

Most analyses of farm policy focus on totals. This article adds another dimension to the debate, often missing. It identifies 702 counties heavily dependent on income from farming and likely to be hardest hit by policy changes. Within that group of counties, those with higher concentrations of farm-related employment are more affected than others.

*** 5.4 ***

Rodefeld, Richard D., Jan Flora, Donald Voth, Isao Fujimoto, and Jim Converse. *Change in Rural America, Causes, Consequences, and Alternatives.* St. Louis: C.V. Mosby Company, 1978. ISBN 0-8016-4145-4.

"Agriculture everywhere is much more organized around the institutions of property than around those of occupation," Stinchcombe observes in his discussion of farm organization structure and the consequences of its change. The major purpose of this collection of readings is to provide the reader with a better understanding of twentieth century changes in the rural sector of the United States. Section three deals specifically with the consequences of changes in agricultural technology for rural neighborhoods and communities. The economic forces stemming from changes in agricultural technology have encouraged declining farm numbers, increasing farm size, more leisure, and greater mobility of rural people. Community decline as measured by population loss, has been at a maximum since declining manpower needs in agriculture occurred simultaneously with forces that acted to relocate and centralize many business and community functions into larger units.

*** 5.5 ***

United States. Department of Agriculture. *Contours of Change.* The 1970 Yearbook of Agriculture. Washington, DC: U.S. Government Printing Office, 1970.

An entire section of this reference volume is devoted to "Country and City--One Nation." Many issues arise as the city moves into the country and the population of isolated rural communities move to the city. The major challenge is to encourage balanced development and maintenance of established rural communities.

The Changing Rural Community

*** 5.6 ***

Carlin, Thomas A., and Linda M. Ghelfi. "Off-Farm Employment and the Farm Sector." *Structure Issues of American Agriculture.* Agricultural Economic Report, no. 438. Washington, DC: Economics, Statistics and Cooperatives Service, U.S. Department of Agriculture, 1979.

The shift toward more off-farm work by farm families is one of the most dramatic changes taking place in U.S. agriculture. Off-farm work is more prevalent among operators of small farms, but some operators of all size units are working off the farm. Those that hold off-farm jobs tend to have more specialized farming operations and use more labor-saving machinery than full-time farmers. Off-farm work is also more common among farm operators less than 45 years old. For many, part-time farming is a way of life.

*** 5.7 ***

Cooperative Extension System. *Revitalizing Rural America.* Washington, DC: Extension Committee on Organization and Policy, National Association of State Universities and Land Grant Colleges, 1986.

The survival of rural America, both the farms and smaller communities, is dependent upon the expansion of income and employment opportunities in rural areas. Staggering declines in farmland values and the instability of net farm income, coupled with increased agricultural competition and the economic decline in other natural resource-based industries, have brought the reality of the rural crisis home. Within the roots of the current crisis lie the seeds of rural revitalization.

*** 5.8 ***

Hustedde, Ron, Ron Shaffer, and Glen Pulver. *Community Economic Analysis: A How to Manual.* Ames, IA: North Central Regional Center for Rural Development, 1984.

This manual is intended for the person interested in the analysis of a community's economy. Numerous tools and techniques can be used to give some insight to the functioning of a community. The manual uses a series of questions to guide a person through the analysis process. The community is intimately linked with the rest of the world through the inflow and outflow of income and goods. And the community uses resources to produce the output it sells. Community economic analysis is a systematic examination of the components of these processes.

*** 5.9 ***

Johnston, Bruce F., and W.C. Clark. *Redesigning Rural Development: A Strategic Perspective.* Baltimore: Johns Hopkins University Press, 1982. ISBN 0801827310, 0801827329 (paperback). LC 81-17138.

An agricultural economist and a systems analyst of resource and environmental management have collaborated on this study. The authors do not claim to offer a cure for the immature field of rural development or an easy solution for development program failures. Instead, they attempt to construct a strategic policy analysis perspective on rural development that can be useful to practitioners and policy makers.

*** 5.10 ***

Jordan, Max, and Tom Hady. "Agriculture and the Changing Structure of the Rural Economy." *Structure Issues of American Agriculture.* Agricultural Economic Report, no. 438. Washington, DC: Economics, Statistics and Cooperatives Service, U.S. Department of Agriculture, 1979.

"Generally, there is little evidence to determine whether encouraging smaller scale farming is a logical strategy for fostering renewed social and economic activity in rural areas," Jordan and Hady conclude. As America has shifted from an agrarian to an industrial-service society, the importance of agriculture as a source of income and employment in rural counties has diminished steadily. Observing national data, the structure of agriculture and the rural community do not seem importantly related. It is easy to overestimate the importance of declining farm numbers in explaining the decline of small rural communities, but they are likely to be important in the decline of some communities.

*** 5.11 ***

Korsching, Peter F., and Judith Gildner, eds. *Interdependencies of Agriculture and Rural Communities in the Twenty-first Century: The North Central Region.* Ames, IA: North Central Regional Center for Rural Development, 1986.

A conference examined the interdependencies of agricultural and rural development, the impact of agricultural development on rural areas, the impacts of community development on agriculture and their implications for policy recommendations for rural areas. With many farm operators facing financial problems, the major question is what is politically feasible in the way of help for rural communities. A belief seems to prevail that agricultural communities must organize themselves to provide the leadership to develop off-farm employment and diversify their economic base.

*** 5.12 ***

Manchester, Alden C. *Agriculture's Links with U.S. and World Economies.* Agriculture Information Bulletin, no. 496. Washington, DC: Economic Research Service, U.S. Department of Agriculture, 1985.

The author concludes that the food and fiber system is one of the largest sectors in the U.S. economy including not only farms but also farm supply stores, machinery dealers, fertilizer distributors, farm input manufacturing plants, and all the activities involved in moving food and fiber from the farm to the dinner table or clothes closet, accounting for nearly 20 percent of the nation's gross national product. The system requires many more nonfarmers than it does farmers. The linkages between the farm economy and the national economy are so close that conditions beyond the farm gate can have as much effect on the well-being of agriculture as farm programs have.

*** 5.13 ***
McGranahan, David A. "Changes in Age and Structure and Rural Community Growth." *Rural Development Perspectives* 1:3 (June 1985):21-25.

Rural community growth is an uneven process. First, the population changes as a result of migration into and out of the community. This change has proven not only uneven but also unpredictable. A second aspect of uneven growth relates to changes in the age structure. The proportion of elderly in the population increased faster in rural communities than in urban areas.

*** 5.14 ***
Milkove, Daniel L., Patrick J. Sullivan, and James J. Mikesell. "Deteriorating Farm Finances Affect Rural Banks and Communities." *Rural Development Perspectives* 2:3 (June 1986):18-21.

Financial problems in the agricultural sector are eventually transmitted to farm lenders. As cash flow problems cause farmers and farm-related businesses to fall behind on loan payments, the quality of lenders' loan portfolios deteriorates. Lenders must set aside reserves to cover actual and anticipated loan losses. These and other adjustments by agricultural lenders to cope with their problem loans can affect credit availability for the community at large.

*** 5.15 ***
Miller, James P. "Rethinking Small Businesses as the Best Way to Create Rural Jobs." *Rural Development Perspectives* 1:2 (February 1985):9-12.

More than half of the new jobs in rural areas are created by branch plants of large corporations. Many areas try to encourage small local firms as sources of new jobs. New data, however, show that such firms create less than a third of new jobs and they are an unreliable employment source because many fail within their first five years of business.

*** 5.16 ***
Organization for Economic Co-Operation and Development. *Rural Public Management*. Paris: OECD, 1986. ISBN 9264128581 (paperback).

Rural areas have been undergoing many, and sometimes rapid, changes during recent years. In many OECD countries, declines in agricultural employment that caused large scale migration out of rural areas have begun to turn around. Population growth has been accompanied by such a diversification and strengthening of the economies in many OECD countries that the likelihood of these population gains being permanent increases.

*** 5.17 ***
Pigg, Kenneth E. *Rural Economic Revitalization, the Cooperative Extension Challenge in the North Central Region*. Ames, IA: North Central Regional Center for Rural Development, 1986.

Many of the rural communities in the North Central Region have economies that are closely tied to farming and therefore suffer economic hardship in direct relationship to those encountered by farmers. However, a dominant characteristic of communities in the region is their diversity and the economic conditions of other communities in the region may be more closely related to changes in the structure of American manufacturing, the aging of the population, or the structural and physical isolation experienced by inadequate access to transportation and communications infrastructure.

*** 5.18 ***
Powers, Ronald C., and Daryl J. Hobbs. "Changing Relationships between Farm and Community." *The Farm and Food System in Transition*, no. 46. East Lansing: Michigan State University. Cooperative Extension Service, 1984.

The connection between agriculture and community is as ancient as farming. Throughout the world, agricultural communities tend to be mirror images of their natural environments and of the kinds of agriculture practiced by their members. Since 1950, farming and agricultural communities have been significantly redefined. Farm consolidation has occurred. Manufacturing has moved into rural communities. Social Security and other pension benefit plans have increased sources of rural income. People living in rural areas are commuting to work some distance away. The increased economic diversity of many rural communities has made it possible for an increasing number of farm families to acquire off-farm income.

*** 5.19 ***
Tweeten, Luther, and George L. Brinkman. *Micropolitan Development, Theory and Practice of Greater Rural Economic Development*. Ames: Iowa State University Press, 1976. ISBN 0813818508. LC 76-163.

The term micropolitan (nonmetropolitan) development originated because economic problems of micropolitan areas differ from metropolitan (large urban) areas. The purpose of the book is to help the student learn the fundamentals of greater-rural development. With an integrated overall perspective and using resources available in this nation, reasonable goals for micropolitan development can be achieved--but not by doing more of what has been done in the past, the authors assert.

*** 5.20 ***
United States. Congress. Joint Economic Committee. *New Dimensions in Rural Policy: Building upon Our*

Heritage. S. Prt. 99-153. 99th Congress, 2d Session. Washington, DC: U.S. Government Printing Office, 1986.

The Joint Economic Committee provides perspectives on trends in employment, agriculture, the farm family, the rural community, small business, education and health, and opportunity for economic development in the rural community. No review of policy for rural America is complete without a reminder of the urgent need to protect and preserve our soil and water, Breimyer emphasizes in his reflections on economic policy.

*** 5.21 ***

United States. Congress. Office of Technology Assessment. "Impacts on Rural Communities." In *Technology, Public Policy, and the Changing Structure of American Agriculture*. OTA-F-285 221-249. Washington, DC: U.S. Government Printing Office, 1986. LC 85-600632.

The impacts of technological and structural change in agriculture do not end with the individuals who live and work on farms. As with individual farmers, some communities are likely to benefit from change, while others are likely to be affected adversely. The Office of Technology Assessment reported a wide range of diversity in the farming, economic structure, and patterns of change in five different regions. The study noted a clear picture of adverse relationships between farm structure and the welfare of rural communities in the California, Arizona, Texas, and Florida counties where the largest farms are established. Large scale industrialized farming in these states is strongly associated with high rates of poverty, substandard housing, and exploitative labor practices. In the coastal zone of the South, the development of large scale, labor-intensive production of fruits, vegetables, and dairy products could also result in making this an area with low farm wages and considerable rural poverty. In the Northeast, rural communities have a low overall dependence on income from farming. In the Midwest, alternative sources of employment in manufacturing and service industries have been relatively pervasive. In the Great Plains and the West, a strong potential for development of a high concentration of farm production exists. Less opportunity for industrializing and large numbers of available hired field workers exist so loss of population and small retail firms could continue.

*** 5.22 ***

United States. Department of Agriculture. *1986 Agricultural Chartbook*. Agricultural Handbook, no. 663. Washington, DC: U.S. Government Printing Office, 1986.

This volume of charts and graphs is a goldmine of current information on farm statistics, population and rural development, consumer affairs, food and nutrition programs, U.S. trade and world production, and commodity trends.

Farming, Employment and Related Businesses

*** 5.23 ***

Batie, Sandra S., and J. Paxton Marshall, eds. *Restructuring Policy for Agriculture: Papers from a Symposium*. Blacksburg: Virginia Polytechnic Institute and State University, College of Agriculture and Life Sciences, 1984.

"The mistakes made in agricultural and food policy are more the result of political knee-jerk reaction than the state of scientific knowledge in the field of agricultural economics," Knutson comments in his discussion of papers developed for this symposium at the annual meeting of the American Association for the Advancement of Science. The papers provide special insights into the consequences of policies on farm families in rural communities, the use of policy to increase access to capital services in rural areas, policy toward alternative farming systems, developing policy to achieve lower rates of soil erosion on fragile lands, and the alternatives to current commodity programs. The editors conclude that restructuring policy would be a complex undertaking, both because national objectives can be in conflict and because maintaining existing programs suits various interests.

*** 5.24 ***

Deavers, Kenneth L., and David L. Brown. "A New Agenda for Rural Policy in the 1980's." *Rural Development Perspectives* 1:1 (October 1984):38-41.

Government policies and programs to aid rural communities need to be reconsidered. Most were adopted in the 1960s and rural conditions changed greatly over the last two decades. Rural areas still need help to continue their momentum and to respond to the new conditions spawned by growth.

*** 5.25 ***

Dillman, Don A., and Daryl J. Hobbs, eds. *Rural Society in the U.S., Issues for the 1980s*. Boulder, CO: Westview Press, 1982. ISBN 0865311005, 086531263X (paperback).

The authors take a serious look at such issues as who will operate our farms and how those farms will meet rising demands for food; how higher energy costs will change life in rural areas; what the current and future needs of rural families and their communities are; who in fact live in these communities; what can be done about escalating rural crime; and recent social changes that have disrupted the traditional patterns of rural society.

The Influence of Changing Agriculture on the Rural Community

*** 5.26 ***

Dorow, Norbert A. "The Farm Structure of the Future: Trends and Issues." *The Farm and Food System in Transition*, no. 17. East Lansing: Michigan State University. Cooperative Extension Service, 1984.

The changing farm structure will continue to affect economic and social conditions in rural areas, but not to the extent that has in the past. Farming no longer dominates rural America in many sections of our country. Farm people make up only 15 percent of the national rural population. If future policies are to help maintain viable rural communities in areas dominated by farming, these policies must help maintain the family farm system, encourage economic development and non-farm job opportunities in rural areas, and help part-time farmers to improve their non-farm incomes.

*** 5.27 ***

Hayes, Michael N., and Alan L. Olmstead. "Farm Size and Community Quality: Arvin and Dinuba Revisited." *American Journal of Agricultural Economics* 66:4 (November 1984):430-436.

Controversy about the economic efficiency and the negative social ramifications of large farms have been a continuing theme. A comparative analysis of Arvin and Dinuba, California, suggests that many factors besides differences in farm size contributed to Arvin's retarded community development. Rather than being closely matched communities, the two towns had developed within significantly different economic, demographic, and geographic settings. Goldschmidt's hypotheses that large farms accounted for differences in community quality may still be correct; but, because of methodological flaws, his study of Arvin and Dinuba offers little support for this assertion.

*** 5.28 ***

Lu, Yao-chi. "Impacts of Technology and Structural Change on the Agricultural Economy, Rural Communities, and the Environment." *American Journal of Agricultural Economics* 67:5 (December 1985):1158-1163.

As agricultural resources become more concentrated in the hands of large operators, sociologists are concerned about the possible adverse impacts on the quality of life in the rural communities. While concentration of resources has no uniformly adverse effects on the quality of life of rural communities across the country, it does have adverse effects in localized areas.

*** 5.29 ***

Moxley, Robert L. "Agriculture, Communities, and Urban Areas." In *New Dimensions in Rural Policy--Building upon Our Heritage*. Joint Economic Committee, U.S. Congress. Washington, DC: U.S. Government Printing Office, 1986.

Moxley summarizes a series of studies that have endeavored to show the relationships of size of farms and the amount of community involvement and development by its local inhabitants. He concludes that size, location, accessibility, and the place of urban areas in a regional hierarchy have some effect on agricultural characteristics. However, he warns that with all macro-level research perspectives, there are limitations to their explanations of the specific mechanisms of influence.

Trade Patterns Related to Farming

*** 5.30 ***

Shaffer, Ron, Priscilla Salant, and William Saupe. "Understanding the Synergistic Link between Rural Communities and Farming." In *New Dimensions in Rural Policy: Building upon Our Heritage*. S. Prt. 99-153. Joint Economic Committee, Congress of the United States. Washington, DC: U.S. Government Printing Office, 1986.

"The current concern about the demise of the family farm and the economic hardships of smaller and medium sized farms is in part stimulated by the distributional consequences of the benefits and burdens of the current economic malaise in farming," the authors assert. This chapter examines interdependence between the rural economy and farming. Even though the farming economy has grown according to many forms of measurement, it no longer dominates rural or nonmetropolitan America.

*** 5.31 ***

Smith, Stephen M. "Diversifying Smalltown Economies with Nonmanufacturing Industries." *Rural Development Perspectives* 2:1 (October 1985):18-22.

The developing economies of small towns and rural areas need more than just factories. Service industries, too, can generate employment and local income. They are usually cleaner than factories and smaller, so they blend in with the local economy rather than dominate it.

Competition and Conflicts over Land Use

*** 5.32 ***

Barrows, Richard L., and David L. Chicoine. "Land for Agriculture." In *Resources, Food and the Future, North Central Regional Extension Publication 222*. Columbus: Ohio State University. Cooperative Extension Service, 1984.

In this paper, along with others in this publication, the authors take a long-range look at the future availability of agricultural land in an urban society. The major theme points out that U.S. cropland is adequate to meet future domestic demand but for a high level of future export demand, the outlook is less certain. They conclude that public policies have a major influence on the land market, how land is used, and who owns it.

*** 5.33 ***
Batie, Sandra S. "Soil Conservation Policy for the Future." *The Farm and Food System in Transition*, no. 23. East Lansing: Michigan State University. Cooperative Extension Service, 1984.

Much uncertainty remains about the importance of soil erosion to future food and fiber production, but it is precisely this uncertainty that makes the case for public concern and public action. Conservation, like housekeeping or yard work, is a continuing process. To be effective, some conservation practices must be maintained on a continuing basis. New conservation efforts are needed despite large expenditures made in the past, Batie concludes.

*** 5.34 ***
Debraal, J. Peter. "Foreign Ownership of U.S. Farmland: What Does It Mean?" *The Farm and Food System in Transition*, no. 34. East Lansing: Michigan State University. Cooperative Extension Service, 1984.

The Agricultural Foreign Investment Disclosure Act of 1978 requires all foreign owners of U.S. agricultural land (farm and forest land) to submit reports detailing among other things, the amount of agricultural acreage they own. The Act also directs the Secretary of Agriculture to determine the effects of such ownership. On December 31, 1983, foreigners reported that they owned 13.7 million acres, slightly more than 1 percent of the 1.29 billion acres of privately owned U.S. agricultural land.

*** 5.35 ***
Debraal, J. Peter, and T. Alexander Majchrowicz. *Foreign Ownership of U.S. Agricultural Land through December 31, 1985*. ERS Staff Report, no. AGES860327. Washington, DC: Economic Research Service, U.S. Department of Agriculture, 1986.

Foreigners owned 12.1 million acres of U.S. agricultural land as of December 31, 1985. This is slightly less than one percent of all privately held agricultural land and 0.5 percent of all land in the United States. These and other findings are based on an analysis of reports submitted in compliance with the Agricultural Foreign Investment Disclosure Act of 1978.

*** 5.36 ***
Shortle, James S., and Donald J. Epp. "The Environmental Connection." *The Farm and Food System in Transition*, no. 37. East Lansing: Michigan State University. Cooperative Extension Service, 1984.

Since the early 1970s the government has launched a number of public programs aimed at improving air and water quality. Agriculture is both a source and a "victim" of environmental degradation. Environmental protection policy is not static. Policy corrections to achieve and maintain improved environmental quality will be forthcoming. Adjustments will also be aimed at improving the efficiency of environmental regulations and achieving an appropriate balance between environmental and other social goals.

*** 5.37 ***
Woodruff, Archibald M., ed. *The Farm and the City, Rivals or Allies?* Englewood Cliffs, NJ: Prentice-Hall, 1980. ISBN 0-13-304980-9.

The patterns of life and the patterns of land use in the United States changed dramatically in the years following World War II. Farms lost population and accounted for less than four percent of the population by the end of the 1970s. Prime farmland was voraciously consumed by suburban real estate development. This book does not try to suggest pat answers but seeks to clarify serious questions about land use and farmland preservation.

Farming and the Demand for Public Services

*** 5.38 ***
Baumel, C. Phillip, Cathy A. Hamlett, and Gregory R. Pautsch. *The Economics of Reducing the County Road System: Three Case Studies in Iowa*. Report DOT/OST/P-34/86-035, Washington, DC: U.S. Department of Transportation, 1986.

The authors examine the economic benefits and costs associated with alternative strategies for abandoning low volume rural highways and bridges. Some findings suggest limited cost savings from abandonment of county roads with no property access in areas of large non-farm rural population; relatively high cost savings from the abandonment of roads with no property access in areas with small rural population; the largest savings are likely from the conversion of public dead-end gravel roads with property or residence accesses to private drives.

*** 5.39 ***
Berg, George L., Jr. "Rural Roads and Bridges." In *New Dimensions in Rural Policy: Building upon Our Heritage*. S. Prt. 99-153. Joint Economic Committee, U.S. Congress. Washington, DC: U.S. Government Printing Office, 1986.

For many decades, the success of American

agriculture has been directly related to the ability of our transportation network to respond to the essential marketing and transportation needs of an expanding agricultural economy. Without an adequate transportation system, rural Americans and the agricultural community would not be able to produce or market the commodities that currently feed and clothe the nation and help reduce the national trade deficit.

*** 5.40 ***

Brohan, Mark. "Tight Times for Rural Governments." *Farmline* (November 1986):8-9.

Many rural governments and school systems are caught in a financial vise of rising costs and falling revenues. The elimination of the federal revenue sharing program, once a key source of money for some counties, cities, and townships, and possible cuts in state aid add to their concerns. Declining farmland values account for much of the reduction in tax revenues.

*** 5.41 ***

Chicoine, David L. "Issues and Implications of the Financial Stress in Agriculture: The State/Local Government Finance Dimension." *Agricultural Finance Review* 47 (March 1987: Special Issue).

Depending on the importance of farming to state and local economies and the structure of public finance and governmental service responsibilities, the economic stress on farms will adversely affect the performance of public revenue systems and the provision of a broad range of local government services. Chicoine identifies these possible implications of farm financial stress: dependence of rural local governments on property taxes, use-value land assessments, decreases in property tax assessments, lower revenue and effects on bond credit ratings, exposure of state governments from lower incomes and sales that would affect tax revenues, and a weakened farm property tax base. Fiscal implications for state and local governments will depend on the structure of the state local tax and expenditure systems.

*** 5.42 ***

Cigler, Beverly A. *Setting Smalltown Research Priorities, the Service Delivery Dimension.* Staff Report, no. AGES 860818. Washington, DC: Economic Research Service, U.S. Department of Agriculture, 1987.

The author examines alternative options for local government service delivery and financing in rural areas with a focus on potential constraints to their use. Research gaps are identified and topics suggested for future research with a focus on service delivery and financing options, policy implementation, and local capacity-building.

*** 5.43 ***

Doeksen, Gerald A., and Janet Peterson. *Critical Issues in the Delivery of Local Government Service in Rural America.* ERS Staff Report, no. AGES860917. Washington, DC: Economic Research Service, U.S. Department of Agriculture, 1987.

Technological changes, an increase in demand for quality community services, and environmental controls have created conditions of continual change in the delivery of rural services. Specific community service topics included in this report are education, water, crime and police protection, fire, hospitals, and solid waste. Some of the research priorities identified are maintenance of current community service facility budgets, expansion of production function research to include a current database, clarification of inputs and outputs, evaluation of alternative proposed policies, and the development of new theories for delivery of community services.

*** 5.44 ***

Reeder, Richard J. *Rural Governments--Raising Revenues and Feeling the Pressure.* Rural Development Research Report, no. 51. Washington, DC: Economic Research Service, U.S. Department of Agriculture, 1985.

Some local governments in nonmetropolitan areas, especially those in the West and in other largely rural areas, experienced during the mid-seventies severe levels of fiscal stress that were associated with high and rising local taxes. These local governments may be forced to cut back their rural development activities in the eighties. The high cost of providing public services in sparsely populated areas contributed substantially to rural fiscal austerity.

*** 5.45 ***

Reeder, Richard J. *Nonmetropolitan Fiscal Indicators: A Review of the Literature.* ERS Staff Report, no. AGES 830908. Washington, DC: Economic Research Service, U.S. Department of Agriculture, 1984.

Fiscal indicators are important to rural development because they identify the nature and severity of local government fiscal stress in nonmetropolitan areas. The author discusses the state of the art of rural fiscal indicators by comparing rural and urban indicator studies. Although they have certain fundamental characteristics in common, rural and urban studies emphasize different types of fiscal difficulties.

*** 5.46 ***

Sofranko, Andrew J., and David L. Chicoine. "Rural Communities in an Urban State." *Illinois Research* 28:3 (Fall 1986):11-13.

The authors point out that the current crisis

has heightened concern for rural social relationships, for support and service sectors linked to farming, and for individuals and families directly implicated in the present crisis. The effects on rural areas will depend on the importance of farming in the local economy.

*** 5.47 ***
Sokolow, Alvin D. "Local Governments in Rural and Small Town America: Diverse Patterns and Common Issues." In *New Dimensions in Rural Policy*. Joint Economic Committee, U.S. Congress. Washington, DC: U.S. Government Printing Office, 1986.

The major issues faced by rural governments in the 1980s include fiscal strain, decrease in general revenue sharing, intergovernmental obligations, obtaining information and expertise to operate efficiently, and maintaining continuity in office. The major challenge is to find the optimal blend of local autonomy and intergovernmental responsibility for small local government units.

*** 5.48 ***
Waldo, Arley D., ed. *Property Taxes...Reform, Relief, Repeal?* North Central Regional Extension Publication, no. 39. rev. ed. St. Paul: University of Minnesota. Cooperative Extension Service, 1980.

With property tax by far the most important tax to finance local government, its operation and administration are important considerations as farmers are faced with financial problems. How property tax relief might be applied or whether property taxes should be repealed remain major issues in many rural areas.

*** 5.49 ***
Walzer, Norman, David L. Chicoine, and Ruth T. McWilliams. "Rebuilding Rural Roads and Bridges." *Rural Development Perspectives* 3:2 (February 1987):15-20.

Half of the rural low-volume roads in four midwestern states need major repair. A fifth of them should be reconstructed. Townships are responsible for maintaining those roads, and financing such large-scale repairs is beyond the capabilities of most small governments. Selective maintenance may be the only way to keep all the roads open.

Financial Stress in Farming and Rural Community

*** 5.50 ***
Dillman, Don A. "Social Issues Impacting Agriculture and Rural Areas as We Approach the 21st Century." In *New Dimensions in Rural Policy: Building upon Our Heritage*. S. Prt. 99-153. Joint Economic Committee, Congress of the United States. Washington, DC: U.S. Government Printing Office, 1986.

Dillman warns that there is reason for concern that rural America might soon become a wasteland of decaying institutions. The farm financial crisis of the 1980s is only the most visible manifestation of a fundamental transition occurring in American society as we shift from a mass society to an information age. The transition poses sociological as well as economic and technological issues that must receive attention, he concludes.

*** 5.51 ***
Heffernan, Judith Bortner, and William D. Heffernan. "When Families Have to Give Up Farming." *Rural Development Perspectives* 2:3 (June 1986):28-31.

The shock of having to give up the family farm affects both the farm family and their community. This study of 40 Missouri farm families shows what they went through in giving up their farms and how they have fared since. Withdrawing from family and friends, experiencing feelings of worthlessness, and maintaining periods of silence present especially difficult problems for families and communities.

*** 5.52 ***
Petrulis, Mindy, Bernal L. Green, Fred Hines, Richard Nolan, and Judith Sommer. *How Is Farm Financial Stress Affecting Rural America?* Agricultural Economic Report, no. 568. Washington, DC: Economic Research Service, U.S. Department of Agriculture, 1987.

The authors have made in-depth comparisons of rural America's current economic health by region and by reliance on farming. Recent farm financial stress has hit the Midwest the hardest because farming there is specialized and the Midwest's overall economy has not rebounded from the 1980-82 recessions. Though the number of farmers working off the farm is growing nationally, more plentiful nonfarm jobs in the Northeast and South have helped offset farmers' financial stress. Economically diversified and densely settled areas with younger residents have fared best, while heavily farming-dependent areas have fared worst and lost population.

General References

*** 5.53 ***
Green, Bernal L., and Thomas A. Carlin. *Agricultural Policy, Rural Counties, and Political Geography*. ERS Staff Report, no. AGED850429. Washington, DC: Economic Research Service, U.S. Department of Agriculture, 1985.

Counties that depend economically on farming and receive large amounts of federal commodity payments will receive the greatest, most direct impact of farm policy adjustments. More than 700 of the nation's 2,443 rural counties depend on farming for at least 20 percent of their income and employment. About a third of the nation's 2.3 million farmers live in these counties.

*** 5.54 ***

Whiting, Larry R., ed. *Communities Left Behind, Alternatives for Development*. Ames, IA: North Central Regional Center for Rural Development, 1974.

Public concern about rural development arises because of the mal-distribution of the costs and benefits that have accompanied both national economic growth and the advance of technology. Communities may decline for various reasons: natural disasters, exhaustion of natural resources, loss of transportation advantages, and loss of political status. Many rural areas of the North Central region that have extensive areas unsuited to agriculture and well-suited to recreation or retirement have been attracting population rapidly.

CHAPTER 6

GOVERNMENT FARM COMMODITY PROGRAMS

Current government farm programs aim to maintain farm income with emphasis on family farms, keep major export commodities competitive in world markets, and limit governmental budget costs within limits that are politically acceptable. Continuing present policies will become increasingly expensive.

"In the rich and mainly industrial countries farmers are paid too much, so they produce too much. In the poor and mainly agricultural countries farmers are paid too little, so they produce too little. Europeans trample Cognac grapes into industrial alcohol; Americans fill Rocky Mountain caverns with butter; Japanese pay eight to ten times the world price for their bowl of rice. Meanwhile, many million Asians and Africans live in rural poverty and go hungry to bed. Do not despair. The mistakes are so large that these contrary policies will soon collapse. Properly staged and handled, that collapse will leave the whole world better off."

The Economist, November 15, 1986.

"...farmers determination to maintain their privileged positions in Asia, Western Europe and the U.S., coupled with a glut of farm commodities, has created such a clash among major powers over farm trade as to threaten the entire free-world trading system..."

The Wall Street Journal, December 4, 1986.

Farm commodity programs have placed the American farm squarely on the horns of an increasingly costly and frustrating dilemma. Due to slow growth of the domestic market, rising productivity has enabled the American farm to supply this market quite adequately and still have increasing amounts of farm commodities available for export. But, in the 1980s, the combined effect of price supports, a strong dollar, global recession, and increasing world food production, was to restrict exports and lock the farm and rural economy solidly in recession.

The 1985 farm bill attempted to make U.S. farm commodities more price-competitive in world trade by sharply lowering price supports on most major export crops; and it aimed to maintain farm income by larger deficiency payments based on high target prices. But the bill worked against farm efficiency by withdrawing more good farm land from production; the budget cost of farm income supports soared above $25 billion a year, far more than had been planned. In years to come, costs may go still higher, adding to the woes of the federal budget deficit and intensifying political pressures for more changes in farm programs.

Current programs are built on traditional goals, which are to (1) maintain farm income, with emphasis on family farms and private ownership of the means of production, (2) keep major export commodities competitive in world markets, and (3) limit governmental budget costs to levels that are politically acceptable. Failures of current policy, especially in the commodity area, prompt us in this chapter to examine some of the programs that have been followed and their consequences. This helps to build a foundation for considering in the next chapter some of the ways for transforming farm policy and programs for the times ahead.

Foundations of Commodity Policy

The current farm commodity programs are almost entirely a product of this century. From early colonial days to the early years of the 1920s, the government was able to avoid direct involvement in farm commodity markets. This era came to an end in the farm depression following World War I. Prices of farm commodities collapsed in the early 1920s, igniting a decade-long agitation for the federal government to come to the aid of the farmer. Some progress was made in helping to ease farm debts, but by far the most important movement centered around a proposal to subsidize exports in order to get higher prices for the major farm commodities such as wheat, corn, cotton, and tobacco. It was thought that such a plan would raise the prices of these products, and other commodities would benefit as well. Bills incorporating an export subsidy to be financed by a processing (sales) tax on the commodity passed Congress twice, but were vetoed by the president. No further action was taken on this proposal, but agitation continued.[1]

In 1929, The Federal Farm Board was established with an initial appropriation of $500 million. These funds would be lent to farm cooperatives to hold certain commodities in storage until given price goals were reached. The board committed its funds, as the law required. Prices collapsed anyway. When the board ran out of money, no more was appropriated. In 1932, the board, in its *Third Annual Report*, its last, concluded that "experience with stabilization...demonstrates that no measure for improving the price of farm products other than increasing the demand of consumers can be effective over a period of years unless it provides more definite control of production than has been achieved so far."[2]

Although a number of other relief measures were debated in Congress and a measure to expand farm credit was adopted, the holdings of the board were liquidated at a loss as farm prices dropped to very low levels. The idea of balancing the farm output with market demand, which had been developed in a book published in 1927,[3] gained more acceptance; and a campaign developed to establish such a policy.[4] Supply control was not far off.

The basic farm policy of the Roosevelt Administration, first set by the Agricultural Adjustment Act of 1933, was to control farm production and marketing to raise prices of major farm commodities, so as to bring agriculture up to a parity with other industries. The act provided for contracts between the secretary of agriculture and individual farmers to control farm production; and contracts between the secretary and processors of farm commodities, called "marketing agreements," to regulate the marketing of certain perishable crops as a means of raising their prices. The act also provided for export or domestic diversion operations, import quotas, relief purchases, and loans to growers.[5]

The act provided for a new agency, called the Agricultural Adjustment Administration (AAA), to be established in the Department of Agriculture under control of the secretary. It also authorized the secretary to establish state and local committees or associations of producers to advise and aid in administration of the act. This was to avoid the charge of "federal bureaucracy" and to secure the help of persons closely informed on local farming conditions in the formulation and administration of adjustment plans.[6]

The general goal of agricultural adjustment was defined in the act as the re-establishment of "prices to farmers at a level that will give agricultural commodities a purchasing power with respect to articles farmers buy, equivalent to the purchasing power of articles in the base period....August 1909-July 1914."

This came to be accepted as a definition for parity, as mentioned in the 1933 act. Although several modifications in regard to the base and methods of computation have been made, and parity prices still have been severely criticized as backward looking and inefficient, the concept of parity for farm commodity prices has tended to survive to the present day.

The central feature of production control was and has continued to be the offering of a specified payment to growers to limit their planted acreage to something less than they planted in a previous base period. This payment, set at so much per bushel for grain, or so much per pound for cotton and tobacco, has generally been high enough to attract the great majority of growers. They in turn have been required to reduce their planted acreage to a percentage of the base, usually in a range of 65 to 90 percent, for example, depending on the program. Specific regulations cover the use of the land that is not planted.

Following passage of the 1933 act, the Commodity Credit Corporation (CCC) was organized to make loans to farmers on commodities held in government-approved storage. If the price of the commodity went above the loan level, the farmer could sell the commodity and use part of the proceeds to retire the loan. If, after the required minimum storage period, the price did not rise to the loan level, the farmer could release the commodity to the CCC, and the government had no further recourse in collecting the loan. This is why these have been called "non-recourse loans." Provisions for establishing marketing quotas were passed in 1934. If producers approved them in a referendum, marketing quotas could limit and control the total amount of a commodity that a producer could market, or sell.

Development of the AAA led to other notable changes in farm policy. The Grazing Act of 1934, sometimes called the Taylor Grazing Act, provided for government regulation of grazing on public lands, a policy that had been started on national forest reserve land in the administration of Theodore Roosevelt more than 30 years before.[7]

In 1935 and 1936, with Congress spurred by the great drought of the mid 1930s, soil conservation was linked with production control. Farmers were paid for shifting specified acreages of soil-depleting crops into soil-conserving crops, the former being largely cash crops such as wheat, cotton, corn, and tobacco, which it was desired to reduce, while the latter were legumes, grasses, and other forage crops. Farmers could receive certain supplemental payments for applying lime or fertilizer on specified acreages. Congress also found it politically attractive to pay farmers to help carry out certain other conservation practices, such as contouring or terracing erodible farm land, building small dams, and sodding waterways.[8] The Tennessee Valley Authority was established as a model for multi-purpose development of a region, involving power, recreational, and farm resources.

The basic dilemma of U.S. farm policy soon began to emerge. In spite of widespread drought in the mid 1930s, good crops starting in 1937 soon brought lower prices. In 1938, Congress set higher levels for price supports and stocks began to accumulate rapidly under the CCC storage programs. By the fall of 1941, before the U.S. entered World War II, stocks of grain were reaching such a high level that the entire system of supply control and price support was threatened with collapse. The outbreak of war postponed the day of reckoning for almost a decade.

Dynamics of Program Operations[9]

During World War II farm prices moved much higher. Although Congress raised price supports far above prewar levels, markets were strong enough to keep carryovers at normal levels, except for cotton. Many articles and books by agricultural economists pointed out that the wartime price supports were far too high to be maintained in peacetime. The economists generally recommended much lower and more flexible supports; or, even better, a return to free-market pricing.[10, 11]

In the post-war period, with massive aid under the European Recovery Act or the Marshall Plan, farm commodities prices boomed, but, in 1949, they dropped sharply. Price supports again came into play. Throughout the 1950s, Congress, under pressure from farm constituents, continued to set price supports for wheat, feed grains, and cotton so high that surpluses accumulated in large volumes.

A number of emergency measures were passed to deal with farm surpluses. The Mutual Security Act of 1951, Public Law 165 (MSA or P.L. 165), and its amendments, especially Public Law 665, provided for food exports as "defense support" items for friendly countries. By 1953, increases in farm surpluses brought pressure for stronger measures. The Agricultural Trade Development and Assistance Act of 1954, Public Law 480, was passed to promote more foreign disposal of U.S. farm surpluses. Whereas the MSA (P.L. 165 and 665) was designed to provide mutual assistance to friendly governments, India and Pakistan especially, with provision for revenue from sales to be used in promoting development of agriculture, Public Law 480 was primarily a domestic bill using the foreign aid program as a means of surplus disposal.

Sales were limited to commodities acquired under price support programs or deemed to be in surplus quantity. Emphasis was on expanding exports with individual countries being permitted to pay in their national currencies, which generally were not convertible into dollars. By 1960, the U.S. was exporting nearly half of its annual wheat crop, with most of it going under the foreign aid programs. Feed grain exports jumped also, but were still only a small proportion of total production. Cotton exports increased, but did not reach the level of the 1920s.

In addition, the Eisenhower Administration sought to avoid more specific production controls by linking land retirement and conservation. The Agricultural Act of 1956 established the Soil Bank, the first large-scale effort since the 1930s to retire land for either production control or conservation. Originally, some 60 million acres of good crop land were intended to be retired under the Soil Bank. The program was funded, however, at less than half this level; and Congressional support for the Soil Bank as a production-control measure weakened. Still, the Agricultural Act of 1958 made price supports mandatory for most feed grains in addition to wheat, cotton, and tobacco. This act also provided an option for corn farmers under which they voted to terminate production-control allotments in 1959, setting the stage for a build-up of the total grain carryover to record levels at the end of the 1959-60 marketing year.

In 1961 the Kennedy administration proposed stronger government intervention. Plans involved more funding for P.L. 480 programs, under a new title "Food for Peace;" more money for food stamps and other domestic relief programs; and more comprehensive and much stronger supply controls, called supply management. These latter would involve stringent penalties for non-compliance by a farmer, such as substantial reduction or withdrawal of all government benefits.

Such a program, it was argued, would save the taxpayers' money. Reduced supply on the domestic market combined with export subsidies would bring higher farm prices. However, in 1963 a referendum on applying this program to wheat was defeated under the leadership of the American Farm Bureau Federation, which strongly opposed mandatory supply controls.[12] However, price supports, implemented through acreage allotments, loans, and storage, were retained.

By the mid 1960s, a series of poor crops in developing countries, accompanied by continuing rapid population growth, resulted in food shortages in a number of countries, creating a sense of shock and an apprehension about the future. Much of the shock was related to successive crop failures in India and other parts of South Asia due to failures of the monsoons and to a poorly developed world trade pattern.

Alarming reports suggested that the world was facing widespread food shortages. *The New Republic* (September 18, 1965), for instance, headed an editorial with the frightening words, "Famine Is Here." A greatly expanded food-aid program was thought to be imperative. James Reston, in the *New York Times* (February 11, 1966), entitled an editorial "Fight 'Em or Feed 'Em." The president's Science Advisory Committee, in a special report, *The World Food Problem* (May 1967), viewed the world food situation as a problem of frightening proportions. It recommended new measures to increase food production in the United States, and a larger U.S. financial program of food aid for the needy countries. For the long term, the committee recommended a stronger commitment to international development programs, and it recognized the need to reduce the rates of population growth in many countries where economic development could not keep pace.

The United States responded by greatly increasing food-aid programs; in one year, for example, it shipped more than 550 million bushels of wheat to India alone, an amount almost equal to the total U.S. wheat consumption in a year. Soon the situation changed again. Two or three good crop years and more abundant rainfall over South Asia coincided with a rapid introduction of new high-yielding wheat varieties and generally improved production practices. The world prospect for food became more heartening.

The United Nations Food and Agriculture Organization (FAO) implied in its *State of Food and Agriculture, 1969*, that the main problem for the future might be managing supplies, rather than overcoming shortages. Production in the developing countries was more than compensating for population growth. The immediate need for food aid dropped off and farm exports continued to be low. In the United States, by the fall of 1970, the index of farm product prices had fallen to its lowest level, relative to non-farm prices, since the desperate days of the 1930s. People worried again about low farm prices, crop surpluses, and the costs of government programs.

Little more than a year later, however, just as these worries were gaining new force, the situation changed again. In 1971, the U.S. trade deficit began to grow at an alarming rate, forcing a devaluation of the dollar in terms of gold. U.S. farm exports became cheaper in terms of other currencies, especially those of Western Europe and Japan. The Soviet Union, plagued by a series of poor crops, began to purchase grain on a large scale. India and Pakistan suffered anew from monsoon failures. The Peru fish catch dropped sharply. Drought followed by typhoons slashed grain crops in the Philippines.

The Peoples Republic of China, also facing poor crops, bought wheat from Canada on a new large scale. The Australian wheat crop was cut by drought. In the broad region of Africa south of the Sahara, called the Sahel, a drought of more than six-years duration brought widespread misery and starvation. By the spring of 1973, food reserves in the United States were drawn down to levels barely sufficient to fill marketing channels. Farm commodity prices, in frenzied trading activity, exploded upward to new unprecedented heights.

Following the period from 1973 to 1975, when grain stocks were reduced to "pipeline" levels and most farm prices were unusually favorable, the 1976 and 1977 wheat crops were large enough to bring a substantial drop in price. Measured in 1978 dollars, the real price of wheat fell from $6.18 per bushel in the 1973 crop year to $2.47 per bushel in the 1977 crop marketing year.[13] This drop in prices caused severe financial problems for many wheat growers. Prices of other grains and soybeans also dropped in real terms, creating more general financial problems in major farm states.

These unfavorable grain prices were the major economic force behind the rise of the American Agriculture Movement (AAM). This militant organization, using disorderly lobbying tactics, such as a highly publicized parade of farm tractors to Washington, claimed credit for securing a rise in the target price for wheat to $3.40 a bushel from $3.00, where it had been set six months earlier by the Food and Agriculture Act of 1977. A new set of production-control measures was introduced to increase the market price.

Then, beginning in 1978, and continuing through 1982, price supports for food grains, mainly wheat, and feed grains were held well above 1977 levels. The total output of grain moved up strongly with total U.S. farm output (see Chapter 2, Table 2-1), but domestic use increased slowly, and exports increased slowly, reaching a peak in 1981. By the end of 1981-82, the total stocks of grain were more than double the 1977-79 average. Thus, the stage was set for a wasteful and unusually expensive carryover of stocks that would be of no direct benefit to farmers or consumers, and which were unneeded to assure that domestic food supplies would be adequate or that the United States would be a reliable supplier of grain on the international market.

The Agriculture and Food Act of 1981 raised price supports for most commodities only modestly higher than they had been set in 1977, and placed increased reliance on direct payments as a means to support farm income. Government costs, however, rose substantially over earlier years. For the 1977 bill, in effect for crop years 1978 through 1981, the total cost was $27.7 billion. For the 1981 farm bill, from 1981 through 1985, total cost was estimated at $63.3 billion.[14]

U.S. farm commodity exports, which reached a peak of $43.8 billion in 1980-81, fell by more than 40 percent by 1985-86, while total world exports of farm commodities grew modestly. U.S. wheat exports, generally the most important component and a leading barometer of farm commodity exports, fell from a peak of 48.8 million metric tons (mmt) in 1981-82 to 32.7 mmt in 1985-86. Correspondingly, wheat exports from all competitor countries increased from 52.5 mmt in 1981-82 to 64.8 mmt in 1985-86. The change in total world wheat trade was not significant, falling slightly from 101.8 mmt in 1981-892 to 97.5 mmt in 1985-86.

The decline in U.S. wheat exports and total farm commodity exports was closely linked to previous loan rates and target prices with the effects compounded by the strong U.S. dollar. Also, a worldwide trend in protectionism for farm commodities and food self-sufficiency was stimulated by the high loan rates and target prices. Markets lost under these circumstances would not be easily regained. The prevailing market crisis added to the difficulty of writing a new farm bill.

The Food Security Act of 1985[15]

The Food Security Act of 1985 was more than two and one-half years in the making. The 1981 farm bill was due to expire at the end of 1985 and new legislation would be required. In May 1983, Senator Roger Jepson (Republican-Iowa), then Chairman of the Joint Economic Committee, convened a series of hearings under the theme "Toward the Next Generation of Farm Policy." In an opening statement, Jepson declared that "farm programs implemented in 1981, 1982 and 1983 have been costly and have proven to be ineffective in reversing the economic deterioration of the farm sector." The 1983 payment-in-kind (PIK) program was cited as a clear indication of the failure of traditional farm programs.

In July 1983, Secretary of Agriculture John R. Block called together a selected group of farm and agribusiness leaders for a Farm Forum to elicit their views on future directions of farm policy. In 1984, these initiatives were followed with a series of hearings in the House Agriculture Committee on long-term farm policy. In 1984, Secretary Block held a series of six meetings across the country to secure more views on farm policy.

Early in 1985, the administration's proposed new farm bill, entitled the Agricultural Adjustment Act of 1985, would have made deep cuts in expenditures on farm programs. It proposed a market orientation in setting loan rates, phasing down target prices by basing them on a percent of the market price, lowering limits on direct payments, phasing out the dairy support program, phasing out direct Farmers Home Administration loans to a loan guarantee program, and banning farm program benefits to any farmer who converted erodible land to crop production that had been idle for ten or more years.

Reactions to these proposals in both the House and Senate were negative, thus setting the stage for a long political dispute in Congress over a new farm bill. Before it was over, a total of 336 witnesses had appeared and filed statements before the House Agriculture Committee and associated subcommittees, and 362 appeared before the Senate Agriculture Committee and its nutrition subcommittee.

The American Farm Bureau Federation, by far the largest of the many farm organizations and generally the most conservative, testified in favor of a more market-oriented policy. Robert

Delano, president of the AFBF stated, "We believe that sound farm policy must be based on letting market forces provide incentives between producers and consumers. It is imperative that farm programs adjust to the signals from the marketplace while giving assistance for orderly marketing, income supplement, production adjustment and market share expansion as we move back to the market system for U.S. agriculture." He called for a stronger and expanded Public Law 480 foreign aid program, an intermediate CCC export credit program, and exemption from the "cargo preference" regulations that require a certain percentage of government-aided exports to be carried in the higher-cost U.S. merchant ships. He also supported target prices as an income supplement, price support loan levels based on a three-to-five year average market price, and a supply-reduction program when carryovers exceed workable storage levels.

Generally, representatives of other farm organizations, such as the National Farmers Union, The American Agriculture Movement, the National Farmers Organization (NFO), and many special-interest commodity organizations, testified in favor of continued government involvement, higher support prices and more effective production controls. From January through June 1985, some 94 bills were introduced in the House and 50 in the Senate, although there was some overlap or duplication of bills in the two chambers. Some 73 amendments were considered, and voted up or down. Hearings continued intermittently from February to mid-September. Committee mark-up sessions and floor debates continued until passage of the House bill in October and the Senate bill near the end of November. Finally, the joint conference committee made the final decisions in meetings that began on December 5 and ended on December 14. The final vote came on December 18. The president signed it on December 23.

The bill, entitled the Food Security Act of 1985, differed greatly from the initial administration proposals, in many ways continuing past policy. Although the decision to set loan rates at 75 to 85 percent of the past three-to-five year average price made U.S. commodities more price competitive, most of the target prices for the major commodities, which determine the levels of deficiency payments, were frozen for two years. The general farm organizations differed on what they asked in regard to target prices, and the final act came close to what the corn and wheat growers wanted--target prices at or near 1985 levels. Price supports for dairy, which have been very expensive, were scheduled to move down in 1987 and 1988 along with a voluntary whole herd buyout program partly financed by producer assessments. Price supports were continued for sugar at 18 cents a pound, three to four times free-market world prices. Other substantial subsidies were continued for peanuts, cotton, honey, rice, wool, and mohair.

Estimates of budget costs were made prior to the time the bill was submitted to the full House and Senate. The costs projected in September 1985, for commodity price and income support only, for the three budget years 1986-88, were $33.8 billion for the House committee bill, and $41.5 billion for the Senate committee bill. By December, after a number of amendments, the Department of Agriculture estimated the three-year cost at $52 to $53 billion. The Congressional Budget Office came up with a figure of $56 billion. Finally, after passage, new estimates by the CBO, in January and February 1986, placed the three-year cost at $64 billion. In late 1986, the actual cost for 1986, reported by the Department of Agriculture, was $25.8 billion. Hence, the three-year cost would be well over $75 billion. Unless the declining dollar provided a strong stimulus for exports, costs might run much higher.

Program benefits would accrue to specific commodity producers in direct proportion to their volume of production. No provisions were made to offer greater benefits to smaller or medium size farmers or to those having the more severe financial problems. The $50,000 limitation on payments to an individual farmer or farm owner, contained in previous acts, was retained, but some new exemptions were added.

More important from the standpoint of equity, the Congress was unable to close off the many types of arrangements large operators and owners use to circumvent the limitations. A large land owner with several tenants, for instance, might evade the limitations by cash-leasing. A large owner-operator might evade the limitations by making family members partners, or by arranging other types of partnerships. Although more specific methods of directing benefits toward smaller or medium size farms were proposed and supported by some farm groups, nothing came of these efforts.

During 1986, farmers, policymakers, and agribusiness leaders criticized the 1985 farm bill severely. American farmers' financial predicaments continued, and often intensified. Record budget costs were a constant worry for Congress and the administration. Efforts to idle more land in succeeding years would push up the already high budget costs. If these efforts were successful in securing higher prices, the added cost would be borne directly by consumers and exports would be discouraged.

For an industry in desperate need of a broader market, the 1985 bill had wide repercussions. In August 1986, representatives of 14 countries, not including the United States and those in the European Community, met in Australia to condemn the subsidies of both the United States and the European Community. They declared that the phasing out and banning of all subsidies that affect international trade in farm commodities should be the number one item on the agenda for future General Agreement on Tariffs and Trade (GATT) negotiations. Although the 14 countries might have been mollified by the fact that U.S. production controls and price supports still tended to restrict U.S. exports, they did not distinguish between the effects of the United States and the European Community subsidy systems. Canada followed this by imposing a new tariff on corn imported from the United States, and by an unprecedented appropriation of one billion Canadian dollars to subsidize their farmers.

The United States, in the meantime, has continued to propose a new round of Multilateral Trade Negotiations (MTN), under the General Agreement on Trade and Tariffs, for the avowed purpose of reducing trade barriers. Since high target prices, however, are in effect an indirect means to subsidize exports, as long as the United States maintains its target prices for farm commodities well above free-market world levels, there is little hope that much progress can be made in reducing trade barriers, or expanding farm commodity exports.

Although the U.S. government can continue programs for land retirement to control production, for price supports, deficiency payments, and export subsidies, these programs will become increasingly expensive and will not achieve the goals usually held for the American farm. In the next chapter we will discuss how farm policy may be transformed for the times ahead.

Endnotes

1. John D. Black. "The McNary-Haugen Movement," *The American Economic Review* XVIII (September 1928). pp.405-427.

2. United States. Federal Farm Board. *Third Annual Report*, 1932. pp.3-4.

3. W. J. Spillman. *Balancing the Farm Output*. New York: Orange Judd Publishing Company, 1927. The plan for balancing the farm output with market demand was developed more completely in John D. Black, *Agricultural Reform in the United States*. New York: McGraw-Hill Book Company, 1929. Chapter 20.

4. William J. Rowley. *M. L. Wilson and the Campaign for the Domestic Allotment*. Lincoln: University of Nebraska Press, 1970. pp.107-141.

5. Edwin G. Nourse, Joseph S. Davis, and John D. Black. *Three Years of the Agricultural Adjustment Administration*. Washington, DC: The Brookings Institution, 1937. pp.78-114.

6. Nourse, et al. op. cit. Chapters III and IX.

7. Wesley Calef. *Private Grazing and Public Lands, Studies of the Local Management of the Taylor Grazing Act*. Chicago: The University of Chicago Press, 1960.

8. Charles M. Hardin. *The Politics of Agriculture, Soil Conservation and the Struggle for Power in Rural America*. Glencoe, IL: The Free Press, 1952, and Robert J. Morgan. *Governing Soil Conservation: Thirty Years of the New Decentralization*. Baltimore: Published for Resources for the Future by the Johns Hopkins Press, 1965.

9. For more extensive discussion of the topics in this section up to 1977, see Harold G. Halcrow. *Food Policy for America*. New York: McGraw-Hill Book Company, 1977. pp.143-332.

10. William H. Nicholls and D. Gale Johnson. "The Farm Price Policy Awards, 1945: A Topical Digest of the Winning Essays," *Journal of Farm Economics* 28 (November 1945). pp.267-283.

11. O.B. Jesness, et al. *Turning the Searchlight on Farm Policy, a Forthright Analysis of Experience, Lessons, Criteria and Recommendations*. Chicago: The Farm Foundation, 1952, and Willard W. Cochrane. *Farm Prices: Myth and Reality*. Minneapolis: The University of Minnesota Press, 1958. Cochrane, who became economic advisor to the secretary of agriculture in 1961, had outlined a comprehensive program for mandatory supply control.

12. Don F. Hadwiger and Ross B. Talbot. *Pressures and Protests: The Kennedy Farm Program and the Wheat Referendum of 1963*. San Francisco: Chandler Publishing Company, 1965.

13. Bruce L. Gardner. *The Governing of Agriculture*. Lawrence: The Regents Press of Kansas, 1981.

14. United States. Economic Research Service. *Financial Characteristics of U.S. Farms, January 1985*. Agricultural Information Bulletin, no. 500. Washington, DC: U.S. Government Printing Office, 1986. p.90.

15. Harold D. Guither. *Tough Choices: Writing the Food Security Act of 1985*. AEI Occasional

Papers. Washington, DC: American Enterprise Institute for Public Policy Research, 1986.

Chapter 6: Annotated Bibliography

Foundations of Commodity Programs

*** 6.1 ***

Benedict, Murray R. *Farm Policies of the United States, 1790-1950: A Study of Their Origins and Development.* New York: The Twentieth Century Fund, 1953. LC 54-7172.

Benedict provides the most comprehensive and detailed study of the origin and development of farm policies up to 1950. He gives the general history of all the major programs including land settlement policy, transportation development, farm credit and monetary policy, and general farm price support and income policy.

*** 6.2 ***

Benedict, Murray R., and Oscar C. Stine. *The Agricultural Commodity Programs, Two Decades of Experience.* New York: The Twentieth Century Fund, 1956. LC 56-12417.

The authors offer the first general summary and detailed analysis of the major farm commodity programs that were begun in the 1930s. They include the major production-control programs for grains and other crops, and analysis of marketing agreements and orders in fruits and vegetables. The authors take a generally limited view of the benefits of the programs.

*** 6.3 ***

Campbell, Christina M. *The Farm Bureau and the New Deal, a Study in the Making of National Policy.* Urbana: The University of Illinois Press, 1962. LC 62-13210.

The author shows the influence of the nation's largest farm organization in making national policy in an important era of farm policy, and some of the interactions of its leaders with their membership and with government leaders. Campbell points out that, because of the Farm Bureau's traditional conservatism, it did not wholeheartedly support all New Deal policies but it did give strong support to the efforts of the government to support prices of major farm commodities.

*** 6.4 ***

Davis, Joseph S. "The McNary-Haugen Plan as Applied to Wheat." In *Wheat Studies of the Food Research Institute.* Palo Alto, CA: Stanford University, 1927.

Davis' study is of continuing importance in evaluating and understanding both the appeal and the weakness of export subsidy proposals. Writing at a time when proposals for export subsidy were dominating debates over policy for the American farm, the author shows the weakness of the export subsidy as a long-term solution to farm income problems.

*** 6.5 ***

Black, John D. *Agricultural Reform in the United States.* New York: McGraw-Hill Book Company, 1929. LC 29-12597.

Black prepared the most comprehensive of the early studies on farm policy in the United States, setting a theme for public action through the early stages of the Great Depression. For instance, he portrayed, in more detail than had been done before, how production controls might be used to raise and stabilize farm prices.

*** 6.6 ***

Fite, Gilbert C. *George N. Peck and the Fight for Farm Parity.* Norman: University of Oklahoma Press, 1954. LC 54-5934.

Fite provides a dramatic account of how the concepts of "parity" and "parity prices" were created, why and how they were accepted as a guiding principle and standard in farm price support programs. Beginning in the 1920s, the concepts were initially defined in terms of "fair exchange value," first explained in a little pamphlet by Peek and Hugh L. Johnson, entitled *Equality for Agriculture* (privately published, 1922). After 1933, "parity" was the accepted term, the term being retained well into the 1960s and 1970s as a guide for price support programs.

*** 6.7 ***

Holland, Forrest. "The Concept and Use of Parity in Agricultural Price and Income Policy." *Agricultural-Food Policy Review.* ERS AFPR-1. 54-61. Washington, DC: Economic Research Service, U.S. Department of Agriculture, 1977.

Holland reviews the historical development of the concept of agricultural parity, discusses its components, and analyzes the economics of its use. Parity price formulas are shown, and the relative importance of commodities in the Index of Prices Paid and Index of Prices Received is detailed. He also discusses the parity ratio as a measure of purchasing power, its shortcomings in accounting for technological change, and other problems related to use of the parity concept.

*** 6.8 ***

Nourse, Edwin G., Joseph S. Davis, and John D. Black. *Three Years of the Agricultural Adjustment Administration.* Washington, DC: The Brookings Institution, 1937. LC 37-4099.

This work concludes a series of studies on the operations of the Agricultural Adjustment Administration established under the Agricultural Adjustment Act of 1933. The authors point out the act's limited achievements. They interpret the actual play of economic forces under the new institutions, define the issues, and set up alternatives, rather than condemn or praise the AAA.

*** 6.9 ***

Rowley, William D. *M.L. Wilson and the Campaign for the Domestic Allotment.* Lincoln: University of Nebraska Press, 1970. ISBN 803207263.

The author presents an accurate and highly informative account of farm policy-making in a critical period of the nation's history. He shows how one person, M.L. Wilson, could animate national policy, and how organizations growing out of such animation can combine to effect important changes in policy.

*** 6.10 ***

Saloutis, Theodore, and John W. Hicks. *Agricultural Discontent in the Middle West, 1900-1939.* Madison: The University of Wisconsin Press, 1951. LC 51-4287.

Two highly talented writers offer a history of the political turbulence engendered by farm-family distress during an important developmental phase of the farm economy in the midwest. The authors show how discontent spawned political movements that were consequential to the entire nation. For instance, without this long period of discontent, which followed close upon the 30-year Populist agitation from about 1865 to 1895, it is unlikely that the major farm programs that have dominated the farm commodity markets since 1933 would have been adopted.

*** 6.11 ***

Rasmussen, Wayne D. "Historical Overview of U.S. Agricultural Policies and Programs." In *Agricultural-Food Policy Review: Commodity Program Perspectives.* Agricultural Economic Report, no. 530. Washington, DC: Economic Research Service, U.S. Department of Agriculture, 1985.

U.S. agricultural policy is as old as the nation itself, Rasmussen explains, going back to the struggle for independence against Great Britain. As the nation developed, policies were implemented to promote exploration and settlement of the frontiers. As the country progressed through cycles of business development, financed at first by farm exports, policies were formulated to help U.S. agriculture keep pace. By the thirties, a national concern arose to improve the depressed farm income situation, which many believed to be the root of the Great Depression. As Rasmussen concludes, "most of our current programs--and much of our current agricultural success--had their origins in this period."

*** 6.12 ***

Spillman, W.T. *Balancing the Farm Output.* New York: Orange Judd Publishing Co., 1927. LC 27-8048.

Spillman was the first to describe a general plan for controlling farm output in the United States. The idea, first publicized in a farm journal a few months earlier, was to induce farmers to organize, at national, state, and local levels, to reduce their production and so to "balance" it with domestic and export demand. Thus the book was a forerunner of the major thrust of U.S. farm commodity programs since 1933.

*** 6.13 ***

United States. Economic Research Service. *History of Agricultural Price-Support and Adjustment Programs, 1933-84.* Agriculture Information Bulletin, no. 485. Washington, DC: U.S. Department of Agriculture, 1984.

Farm commodity price support and production adjustment programs are carried out under a series of interrelated laws passed by Congress. Beginning with the major proposals of the 1920s for handling and marketing farm surpluses, this history records the establishment of price support and adjustment programs under the Federal Farm Board in 1929 and the Agricultural Adjustment Acts of 1933 and 1938. It then traces their evolution through 1984, concluding that "this half century of development is important because it forms the foundation for implementing current and future farm legislation."

Dynamics of Program Operations

*** 6.14 ***

Bredahl, Maury E., Williams H. Meyers, and Keith J. Collins. "The Elasticity of Foreign Demand for U.S. Agricultural Products: The Importance of Price Transmission Elasticity." *American Journal of Agricultural Economics* 61:1 (February 1969):58-63.

The authors contend that a key question "in evaluating the elasticity of export demand is the size of the adjustment of foreign prices to U.S. prices." In many major importing and exporting regions, internal price is largely insulated from U.S. (and/or world market) prices, and these regions do not contribute significantly to the elasticity of U.S. export demand. The authors conclude that "consistently empirical estimates of the elasticity of export demand for specific agricultural commodities are either inelastic or slightly greater than one....These estimates may simply reflect the trade restricting behavior of the real world."

*** 6.15 ***

Chavas, Jean-Paul, and Richard M. Klemme. "Aggregate Milk Supply Response and Investment Behavior on U.S. Dairy Farms." *American Journal of Agricultural Economics* 68:1 (February 1986):55-66.

The authors present a dynamic model of herd composition and supply response in the U.S. dairy sector. By incorporating biological information, they analyze the influences of the economic environment on culling rates and the dynamics of the dairy cow population. Short-run and long-run supply elasticities provide considerable insights on milk supply adjustments over time.

*** 6.16 ***

Chowdhury, Ashok, and Earl Heady. "An Analysis of a Marginal Versus the Conventional Land Set-Aside Program." *North Central Journal of Agricultural Economics* 2:2 (July 1980):95-106.

The authors show that "excess supplies of farm commodities and the resulting decrease in returns to production factors have been persistent problems in American agriculture over much of the last five decades." Two alternatives in acreage diversion directed toward controlling supply to attain higher price levels are analyzed. One assumes an equal percentage set-aside applied to all regions and the other allows the set-aside to be concentrated by regions according to comparative advantage.

*** 6.17 ***

Committee for Economic Development. *An Adaptive Program for Agriculture.* New York: Committee for Economic Development, 1962. LC 62-1945.

The committee, in one of the most important early post-war studies, argued for free markets for farm commodities instead of production controls with high farm price supports. The authors, a committee of prominent economists, recommended a more growth-oriented program adapted to an expanding national and world economy.

*** 6.18 ***

United States. Comptroller General. *1983 Payment-In-Kind Program Overview: Its Design, Impact, and Cost.* Report to the Congress, GAO/RCED-85-89. General Accounting Office. Washington, DC: U.S. Government Printing Office, 1985.

The GAO discusses the Department of Agriculture's 1983 Payment-In-Kind (PIK) program. Being the final GAO work on the program, it ties together the results of several GAO reports, and provides additional analysis on selected aspects of the program. Under PIK, the department gave farmers commodities, instead of cash, to remove cropland from production. GAO found that "PIK cost about $10 billion, reduced farm production and surplus stock levels, and increased farmer net cash incomes." The program was not renewed in 1985.

*** 6.19 ***

Ericksen, Milton H., and Keith Collins. "Effectiveness of Acreage Reduction Programs." In *Agricultural-Food Policy Review: Commodity Program Perspectives.* Agricultural Economic Report, no. 530. Washington, DC: Economic Research Service, U.S. Department of Agriculture, 1985.

Acreage reduction programs are used to reduce production and boost prices. In 1983, they were used to idle a third of the land used for program crops. Ericksen and Collins contend that "acreage reduction programs have been costly and inefficient. Their effectiveness is offset by increased plantings on unrestricted acres. Idling of lower-yielding land reduces their impact on production. Most farmers are paid more than the minimum they would accept to idle the land. Erosive lands generally are not idled. Downward adjustment in land prices as a consequence of technical change is blunted." They conclude that "acreage reduction programs need to be evaluated in relation to these inefficiencies, foreign acreage response, and the production signals given producers by other program provisions."

*** 6.20 ***

Gardiner, Walter H., and Praveen M. Dixit. *Price Elasticity of Export Demand.* Foreign Agricultural Economic Report, no. 228. Washington, DC: Economic Research Service, U.S. Department of Agriculture, 1987.

Gardiner and Dixit define price elasticity of export demand, examine how the adjustment period, government policies, domestic structure of trading countries, and export market shares affect elasticity; they present recent estimates of the elasticity for selected agricultural commodities. They conclude that "although definitive estimates for export demand elasticities are not available, many studies indicate that the short run export demand for U.S. grains, soybeans, and cotton is price inelastic. The long run evidence is even less conclusive. There are substantially fewer studies that estimate the long run price elasticity of demand for U.S. agricultural products, and the estimates vary widely."

*** 6.21 ***

Gardner, Bruce L. "Determinants of Supply Elasticity in Interdependent Markets." *American Journal of Agricultural Economics* 61:3 (August 1979):463-

475.

Gardner elaborates on the implications of equilibrium in a two-product, two-factor model for elasticity of product supply, which is found to depend upon input supply elasticities, alternative product demand elasticity, elasticity of substitution between production inputs, relative factor intensity of the product, and relative importance of the product in its use of resources. He concludes that "these factors interact in a complex manner to determine supply elasticity." He then discusses related approaches and considers several simplified examples in an attempt to provide an intuitive grasp of the workings of the model.

*** 6.22 ***

Gardner, Bruce L. "Robust Stabilization Policies for International Commodity Agreements." *The American Economic Review* 69:2 (May 1979):169-172.

Gardner concludes that "while socially optimum stockpiling rules tend to be fairly robust,...it is important to exercise care in the choice of a buffer stock regime. A striking result of the experiments is that the upper limit on the buffer stock tends to be as important as the price band used. The economic reason is that too large a stock can easily run up storage costs which at the margin provide no corresponding benefits, while a small maximum size leaves the system open to large external costs from shortfalls."

*** 6.23 ***

Groenewegen, J.R., and W.W. Cochrane. "A Proposal to Further Increase the Stability of the American Grain Sector," *American Journal of Agricultural Economics* 62:4 (November 1980):806-811.

The viability of U.S. food and agriculture is dependent on exports to the world grain market. Insulating policies by importers and variable import demands contribute to instability of world grain markets. The authors propose that to reduce vulnerability to this instability, "American food and agriculture policy should move in the direction of insulating domestic food and agriculture from world instability and at the same time maintain its position as a dependable supplier of world grains."

*** 6.24 ***

Hadwiger, Don F., and Ross B. Talbot. *Pressures and Protests: The Kennedy Farm Program and the Wheat Referendum of 1963*. San Francisco: Chandler Publishing Company, 1965. LC 64-8160.

The American Farm Bureau Federation defeated the wheat referendum of 1963, which was the key proposal of the Kennedy administration's policy to control production on the American farm. The administration supported stronger production controls as a means of raising prices of farm commodities while minimizing government expenditures. The Farm Bureau fought the program as an undesirable intrusion of government into the traditional freedoms of American farm families.

*** 6.25 ***

Heien, Dale M. "The Structure of Food Demand: Interrelatedness and Quality." *American Journal of Agricultural Economics* 64:2 (May 1982):213-221.

Heien presents a new empirical demand system that mitigates the effects of multi-collinearity while facilitating the measurement of interrelatedness. He uses duality theory to derive inverse demand functions. On the basis of fourteen food items, he estimates a complete set of demand equations in both quantity and price dependent forms. Heien uses the model to test for the theoretical restrictions of homogeneity, additivity and negativity, as well as the habit formation hypothesis.

*** 6.26 ***

Infanger, Craig L., William C. Bailey, and David R. Dyer. "Agricultural Policy in Austerity: The Making of the 1981 Farm Bill." *American Journal of Agricultural Economics* 65:1 (February 1983):1-9.

The authors examine the legislative actions leading to the creation of the Agriculture and Food Act of 1981 and analyze the influence of the budget process on those legislative actions. They conclude that "the budget process represents another substantial erosion in the ability of agricultural interests to control the farm and food agenda in the 1980s."

*** 6.27 ***

Jesness, O.B., et al. *Turning the Searchlight on Farm Policy, a Forthright Analysis of Experience, Lessons, Criteria and Recommendations*. Chicago: The Farm Foundation, 1952. LC 52-2622.

Thirteen well-known agricultural economists, invited by the Farm Foundation, considered a general farm policy for the post-war era. They recommended "a free market-clearing price policy, ready and equal access to credit for farmers, managerial choices to be free from government intervention, direct supplementary income payments to farmers in depression, and a separate program of adjustment for noncommercial farmers to promote economic equality for agriculture."

*** 6.28 ***

Kinnucan, Henry, and Olan D. Forker. "Seasonality in the Consumer Response to Milk Advertising and Implications for Milk Promotion Policy." *American Journal of Agricultural Economics* 68:3 (August 1986):562-571.

Kinnucan and Forker attempted to measure

the effect of fluid milk advertising on milk sales in New York City. They concluded that "the allocation of generic advertising dollars according to optimization rules during the period 1979-81 would have resulted in a 9 percent increase in returns to dairymen supplying the New York City market. Harmonic variables are used to account for seasonality, a Pascal distribution is used to account for the decay structure, and goodwill elasticities are estimated to indicate the impact of generic advertising on sales."

*** 6.29 ***
Kirkland, Jack J., and Ron C. Mittelhammer. "A Nonlinear Programming Analysis of Production Response to Multiple Component Milk Pricing." *American Journal of Agricultural Economics* 68:1 (February 1986):44-54.

Feeding Holstein cows to maximize revenue over feed costs was analyzed under various scenarios concerning milk component prices. Kirkland and Mittelhammer concluded as a result of their study, "that production response of milk components to component price incentives was inflexible. When compared to baseline solutions generated under a standard federal order pricing system, fat production was within 3.5%, and solids nonfat and fluid carrier within 1.0% of baseline levels for a 100-cow herd. Component pricing scenarios having blend prices equal to the blend price in the baseline scenario were associated with reductions in revenue over feed costs of less than 4%."

*** 6.30 ***
Konandres, Panos A., and Andrew Schmitz. "Welfare Implications of Grain Price Stabilization: Some Empirical Evidence for the United States." *American Journal of Agricultural Economics* 60:1 (February 1978):74-84.

Konandres and Schmitz concluded as a result of their empirical study that, "although United States producers and consumers taken together benefit from policies which would stabilize feed grain prices, this is likely not the case for wheat. The model specifies a U.S. domestic demand relationship for food and feed use, a stock relationship, and a foreign demand sector; these are estimated by ordinary and two-stage least squares methods. The key to the analysis is in testing a well-known theoretical model in which the desirability of price stabilization largely depends on the source of instability i.e., whether instability is generated abroad or is created internally."

*** 6.31 ***
Kramer, Randall A., and Rulon D. Pope. "Participation in Farm Commodity Programs: A Stochastic Dominance Analysis." *American Journal of Agricultural Economics* 63:1 (February 1981):119-128.

The net benefits of participation in farm commodity programs are analyzed with a normative risk model based on stochastic dominance theory. Utilizing entire probability distributions of participation and nonparticipation net returns, the impacts of alternative program features and farm size are investigated. Small changes in program parameters are found to affect participation decisions. It also is demonstrated that farm size can influence participation choices.

*** 6.32 ***
Knutson, Ronald D., James W. Richardson, Danny K. Klinefelter, Mechel S. Paggi, and Edward G. Smith. *Policy Tools for U.S. Agriculture*. College Station: Department of Agricultural Economics, Texas A&M University, 1986.

The authors briefly describe individual policy tools that are most directly related to agriculture and the United States Department of Agriculture. They designed the report "to be a comprehensive list of those policy tools that are used currently, have been used in the past, are used in other countries, or have been proposed for use in the United States. The tools are divided into four general categories: domestic farm programs--designed to raise or stabilize farm prices and incomes; international trade policies--designed to create a more favorable trading environment for U.S. farm products; marketing programs designed to improve farmers' positions in domestic and foreign markets; and credit programs--designed to assure agriculture an adequate supply of debt capital at a reasonable cost."

*** 6.33 ***
LaFrance, Jeffrey T. "The Structure of Constant Elasticity Demand Models." *American Journal of Agricultural Economics* 68:3 (August 1986): 543-552.

LaFrance analyzes the structure of incomplete systems of constant elasticity demand functions and demonstrates that there is a duality theory for incomplete demand systems that is analogous to the duality theory for complete systems. The author concludes that this "theory permits the recovery of that portion of the direct and indirect preferences pertaining to the goods of interest, and we can calculate exact welfare measures for changes in income and in the prices of these goods. For an incomplete system of constant elasticity demands, the Slutsky symmetry restrictions for integratability are presented and the implied structure of the direct and indirect preferences with respect to the prices and goods of interest is derived."

*** 6.34 ***

Langley, James A., Robert D. Reinsel, John A. Craven, James A. Zellner, and Frederick J. Nelson. "Commodity Price and Income Support Policies in Perspective." In *Agricultural-Food Policy Review: Commodity Program Perspectives.* Agricultural Economic Report, no. 530. Washington, DC: Economic Research Service, U.S. Department of Agriculture, 1985.

The authors examine the objectives, performance, effects, and interaction of non-recourse loans, government and farmer-owned stock management activities, and target prices and deficiency payments. They show that "setting loan rates above market-clearing prices increases farm income more than would loan rates used solely for price stabilization. However, relatively high loan rates also increase government stocks, reduce the quantity of domestic and export demand, and increase program costs and food prices. Using the farmer-owned reserve to support farm income has often led to large stock accumulation. Target prices are intended to separate income support from price stability objectives, but deficiency payments also compensate farmers for reducing acreage."

*** 6.35 ***

Leibenluft, Robert F. *Competition in Farm Inputs: An Examination of Four Industries.* Washington, DC: Federal Trade Commission, 1981.

Farm input industries account for more than 40 percent of the total farm production costs in the United States. Imperfect competition is common in all of the markets, with firms generally facing downward-sloping demand curves, or a kinked demand curve characteristic of oligopoly (few sellers). Firms generally compete in terms of brand or company name, and advertise for purposes of product differentiation. Price rigidity is characteristic among the sellers; but this study, which also cites a large number of other studies, did not find any evidence of collusion or practices in restraint of trade.

*** 6.36 ***

Lin, William, Joseph Glauber, Linwood Hoffman, Keith Collins, and Sam Evans. *The Farmer-Owned Reserve Release Mechanism and State Grain Prices.* ERS Staff Report, no. AGES850717. Washington, DC: Economic Research Service, U.S. Department of Agriculture, 1985.

The authors quantify relationships between reserve activities and state grain prices for corn, sorghum, and wheat, and discuss the farmer-owned reserve release mechanisms and some alternatives to the current release mechanism. They conclude that "release of farmer-owned reserve stocks had little or no measurable effect on lower state-U.S. monthly grain price differentials for most of the states studied. The 5-day average adjusted prices based on a production-weighted average and a reserve-weighted average were shown to differ from the price series based on a simple average (the current method) only by a few cents. Setting release prices in each state by adjusting the national release price by the normal state-U.S. grain price differentials would narrow the differential for states where an abnormally wide differential has been the case during release status."

*** 6.37 ***

McCalla, Alex F., T. Kelley White, and Kenneth Clayton. *Embargoes, Surplus Disposal, and U.S. Agriculture: A Summary.* Agriculture Information Bulletin, no. 503. Washington, DC: Economic Research Service, U.S. Department of Agriculture, 1986.

The authors have fulfilled a congressional mandate contained in the 1985 Supplemental Appropriations Bill directing the Economic Research Service (ERS) to conduct "...a study to determine the losses suffered by U.S. farm producers during the last decade as a result of embargoes and the failure to offer for sale on world markets commodities surplus to domestic needs at competitive prices." ERS enlisted the best academic authorities in a joint research effort to produce this study. The International Agricultural Trade Research Consortium, of which ERS is a member and sponsor, was used to identify and solicit participation of university faculty who are experts in international trade. The study, then, is the product of a team of agricultural economists from ERS, 14 universities, and one private research institution. The conclusion was that "embargoes did not cause the farm crisis of the 1980s, and an aggressive export subsidy program to reduce surplus commodity stocks would not have prevented it. The cause more likely rests with radical changes in such worldwide economic conditions as recession, high interest rates, and the value of the dollar. A general export subsidy to dispose of stocks would be more expensive than existing programs although farm income would remain basically unchanged and world price variability would increase."

*** 6.38 ***

Masson, Robert T., and Philip M. Eisenstat. "Welfare Impacts of Milk Orders and the Anti-Trust Immunities for Cooperatives." *American Journal of Agricultural Economics* 62:2 (May 1980):270-278.

Mergers in the late 1960s and early 1970s led to large milk-cooperative monopolies. These monopolies have been able to manipulate federal regulation to advantage over competing farmers and regulation continues to protect established power. A social cost analysis model shows that this power cost society at least $70 million a year prior to antitrust action. The authors conclude that "the antitrust market share standards used to challenge corporate mergers or joint ventures are too stringent for evaluating the anticompetitive impact of cooperative mergers or marketing agreements. But like corporations, cooperatives must be prevented from merger or joint venture where 'the effect may be substantially to lessen competition' if society is to be protected from power which far exceeds that originally intended by the Capper-Volstead Act."

*** 6.39 ***
Paarlberg, Philip L., and Philip C. Abbott. "Oligopolistic Behavior by Public Agencies in International Trade: The World Wheat Market." *American Journal of Agricultural Economics* 68:3 (August 1986):528-542.

A model of the world wheat market treats public policies as endogenous. The oligopolistic nature of international wheat trade is captured by assuming policy makers form conjectures on the slope of the excess demand function they face and use that information to determine domestic and trade policies. The policies reflect differing influences of political interest groups. A U.S. crop shortfall scenario illustrates the different results with endogenous policies compared to the traditional model.

*** 6.40 ***
Paarlberg, Philip L., Alan J. Webb, John C. Dunmore, and J. Larry Deaton. "The United States Competitive Position in World Commodity Trade." In *Agricultural-Food Policy Review: Commodity Program Perspectives*. Agricultural Economic Report, no. 530. Washington, DC: Economic Research Service, U.S. Department of Agriculture, 1985.

The authors concluded that "the decline in U.S. agricultural exports and the U.S. share of world markets since the late seventies, as well as the adjustments presently occurring in U.S. agriculture--lower incomes and lower land values--are not due to the United States becoming a high-cost producer. Rather, they are due to a decline in relative prices of agricultural commodities caused by U.S. and foreign agricultural policies, a rising dollar, the global recession, and debt problems in some importing countries. U.S. farmers remain low-cost producers."

*** 6.41 ***
Quizon, Jaime B., and Hans P. Binswanger. "Income Distribution in Agriculture: A Unified Approach." *American Journal of Agricultural Economics* 65:3 (August 1983):526-538.

Quizon and Binswanger develop a class of partial equilibrium models for analyzing a variety of exogenous shocks in agriculture. The shocks include technical change, changes in the supply of agricultural inputs or in the demand for agricultural outputs, and government interventions. The authors conclude that "the models are capable of tracing income distribution effects of exogenous shocks in agriculture, such as technical change (neutral or biased), changes in factor supply, changes in the final demand, or interregional factor mobility. However, because they are equilibrium models, they are not capable of tracing the fourth class of effects, namely, disequilibrium phenomena, such as the differential adoption of technical changes among farmers in a particular region."

*** 6.42 ***
Sarris, Alexander H., and John Freebairn. "Endogenous Price Policies and International Wheat Prices." *American Journal of Agricultural Economics* 65:2 (May 1983):214-224.

International prices are modeled as Cournot equilibrium interactions of national excess demand functions, which, in turn, are solutions of domestic welfare optimization problems. The authors show "that interaction of national policies can lead to excess depression and instability of both international and domestic prices." The model indicates that under free trade average world wheat prices would be much higher while price instability would be much lower. The European Economic Community policies alone contribute about 80 percent of current price distortions.

*** 6.43 ***
Senauer, Ben, and Nathan Young. "The Impact of Food Stamps on Food Expenditures: Rejection of the Traditional Model." *American Journal of Agricultural Economics* 68:1 (February 1986):37-43.

Senauer and Young conclude that "for food stamp recipients whose normal food purchases exceed their coupon allotment, the traditional economic model predicts that the impact of food stamps on food spending will be the same as for an equal cash transfer." The Tobit analysis in this study indicates that for these recipients, food stamps have a substantially greater impact on at-home food expenditures than an equal amount of cash income. These results reject the traditional model. Several possible explanations of this behavior are discussed.

*** 6.44 ***
Thurman, Walter N. "The Poultry Market: Demand Stability and Industry Structure." *American Journal of Agricultural Economics* 69:1 (February 1987):30-37.

Thurman explores the stability of demand

for poultry meat and discusses specification issues in its estimation. He concludes "that the demand for poultry meat shifted out in the early 1970s. At the same time, the demand relationship between poultry and pork changed from substitution to independence. A second conclusion is that poultry price is predetermined for demand in annual U.S. data, while quantity is not. A predetermined price suggests several market structures. Exogeneity tests are performed to distinguish among them. The results are consistent with a competitive, constant returns-to-scale-industry facing elastic factor supplies."

*** 6.45 ***

Woods, Mike, Luther Tweeten, Daryll E. Ray, and Greg Parvin. "Statistical Tests of the Hypothesis of Reversible Agricultural Supply." *North Central Journal of Agricultural Economics* 3:1 (January 1981):13-19.

The authors estimate aggregate supply equations using four alternative methods to segment the output price variable to measure short-run response of farm output to increasing and decreasing prices. They concluded that the "Null hypotheses of equal price response coefficients for falling and rising prices could not be rejected. The four alternative methods used to measure the response to rising and falling prices provided no consistent evidence to reject the null hypothesis of no difference in response. No method of segmentation used appeared to be clearly superior to the others, but shortcomings in application of some of the methods were apparent....Supply response is low in the short run, both for rising *and* falling prices."

*** 6.46 ***

Yonkers, Robert D., Karen Dvorak, Ronald D. Knutson, Charles W. Bausell, Jr., and Jay R. Cherlow. "Impact of the Milk Diversion Program on Milk Supplies." *North Central Journal of Agricultural Economics* 9:1 (January 1987):107-112.

The authors found that the total Milk Diversion Program contracts equaled 6.9 percent of 1982 national marketings. Then they used regression techniques to quantify the effects of the program on milk production across 33 major milk producing states during the period from January 1984 to March 1985. Their conclusion was that results indicate problems with short-term, voluntary programs such as this one, which did not utilize price as the main incentive for production adjustment.

*** 6.47 ***

Zulauf, Carl R., Harold D. Guither, and Dennis R. Henderson. "Government and Agriculture: Views of Agribusiness and Farm Operators Concerning Selected Issues of the 1985 Farm Bill Debate." *North Central Journal of Agricultural Economics* 9:1 (January 1987):85-98.

During March 1984, the authors surveyed Illinois and Ohio farm and agribusiness operators about farm policy issues. Agribusiness operators expressed significantly less support than farmers for deficiency payments, acreage diversion payments, farmer-owned reserve, and payment-in-kind. On numerous issues, however, the two groups exhibited general agreement. Farmers with sales over $200,000 expressed significantly more support for existing programs than did operators of smaller farms. The authors concluded that "Economic self-interest appears to partially explain differences that exist among and between farm and agribusiness operators. *Food Security Act of 1985* provisions more closely reflect views of farmers who operate medium/large than small/medium farms."

The Food Security Act of 1985

*** 6.48 ***

Gardner, Bruce L., ed. *U.S. Agricultural Policy: The 1985 Farm Legislation.* Washington, DC: American Enterprise Institute for Public Policy Research, 1985. ISBN 0-8447-2256-1.

These papers presented at a conference in January 1985 assess the state of U.S. farming, causes of current difficulties, and prospects for improving matters by reforms in federal farm programs. Major themes are that current problems are strongly related to such macroeconomic factors as federal deficits, inflation, interest rates, and exchange rates; that the extensive integration of U.S. agriculture with the world economy has fundamentally altered the market for farm commodities and placed new constraints on farm policy; and that it is more crucial than ever to have farm commodity programs consistent with market realities.

*** 6.49 ***

Glaser, Lewrene K. *Provision of the Food Security Act of 1985.* Agriculture Information Bulletin, no. 498. Washington, DC: Economic Research Service. U.S. Department of Agriculture, 1986.

The author describes the provisions of the 1985 act for dairy, wool and mohair, wheat, feed grains, cotton, rice, peanuts, soybeans, sugar; income and price supports, disaster payments, and acreage reductions; other general commodity provisions; trade; conservation, research, extension, and teaching; food stamps; and marketing. The report also compares provisions of the 1985 Act with earlier legislation.

*** 6.50 ***

Guither, Harold D. *Tough Choices: Writing the Food*

Security Act of 1985. AEI Occasional Papers. Washington, DC: American Enterprise Institute, 1986.

Not since the major farm legislation of the 1930s has Congress worked so long and hard, or been watched with such anticipation, as it did in developing the Food Security Act of 1985. The committees started with high hopes of making major changes in commodity programs, but ended by generally following the lobbying positions of the major farm commodity organizations. Major changes were made in conservation programs, the formula for setting loan rates was changed to permit loan rates to be lowered while deficiency payments were increased as a result, and more money was authorized to subsidize exports. Guither concluded that "Writing this act was one of the most complex tasks Congress and the Executive branch have ever encountered." The result was a budget-busting compromise.

*** 6.51 ***

Paarlberg, Philip L., Alan J. Webb, Arthur Morey, and Jerry A. Sharples. *Impacts of Policy on U.S. Agricultural Trade*. ERS Staff Report, no. AGES840802. Washington, DC: Economic Research Service, U.S. Department of Agriculture, 1984.

The authors examine the agricultural trade policy environment of the 1985 farm legislation. The report has several central themes. First, U.S. domestic farm programs project a trade policy to other nations; that is, the trade effects of U.S. policies are not neutral. Second, the trade effects of U.S. policies differ according to the export demand situation facing the United States. Third, macroeconomic policies could have a major impact on achieving the objectives of U.S. farm policy established in the 1985 farm legislation.

*** 6.52 ***

Rausser, Gordon C., and Kenneth R. Farrell, eds. *Alternative Agricultural and Food Policies and the 1985 Farm Bill*. Berkeley: Giannini Foundation of Agricultural Economics, University of California, and the National Center for Food and Agricultural Policy, Resources for the Future. Washington, DC: 1984. LC 85-229709.

Economists in the hope of imparting a new sense of direction for the 1985 farm bill organized an important conference on the subject. The contributors, in order of appearance, include such well-known names in the policy area as D. Gale Johnson, Bill Lesher, John Schnittker, Bruce Gardner, Lynn Daft, Willard Cochrane, Luther Tweeten, Alex McCalla, J.B. Penn, Ed Schuh, and Ron Knutson. In addition to these, conference organizers drew on distinguished members of the Berkeley and Davis faculties for substantive papers and comments. The first third of the volume is weighted heavily with criticisms of past price policies with the bias of most authors clearly toward a greater degree of "market orientation" in agriculture. The middle section discusses programs for cotton, tobacco and peanuts, sugar and milk, and includes an empirical study of the effects of increased instability in grain prices on the level and stability of returns to livestock feeders. The final section consists of a miscellaneous group of papers dealing with structural issues, macroeconomic variables affecting agriculture, targeting of soil conservation programs, and the impact of price supports and food subsidy programs on consumers.

*** 6.53 ***

United States. Economic Research Service. *Agricultural-Food Policy Review: Commodity Program Perspectives*. Agricultural Economic Report, no. 530. Washington, DC: U.S. Department of Agriculture, 1985.

Background information for evaluating commodity programs has been brought together in this useful review. Articles provide a historical overview of U.S. farm policies, a description of the general economic setting in which 1985 farm legislation will operate, an evaluation of the performance of current commodity programs, and a discussion of possible alternative policy tools and concepts. Special attention is given to the purpose of commodity programs and an economic assessment of their performance.

*** 6.54 ***

United States. Economic Research Service. *Barley: Background for 1985 Farm Legislation*. Agriculture Information Bulletin, no. 477. Washington, DC: U.S. Department of Agriculture, 1984.

Barley, the third leading feed grain, is grown mainly in the Northern Plains and Pacific regions, and issued mainly for livestock feed and manufacture of alcoholic beverages. Feed use often accounts for more than half of total use. Barley is the most important grain product used by brewers. Exports are small and highly variable. Barley yields have steadily risen, but production costs have also increased. Government loan rates and target prices for barley are based on those for corn. The report concludes that "Government payments to barley growers, while relatively small, have been a significant portion of net returns in some years."

*** 6.55 ***

United States. Economic Research Service. *Corn: Background for 1985 Farm Legislation*. Agriculture Information Bulletin, no. 471. Washington, DC: U.S. Department of Agriculture, 1984.

Corn is the leading U.S. crop, both in volume

and in value. Rising corn yields and market prices strengthened corn farmers' cash flow positions in the late 1970s; but, in the 1980s, loan rates above market-clearing levels and a continued strong U.S. dollar have contributed to a decline in U.S. corn and coarse grain exports. The report concludes that "Higher feed grain prices stemming from production-control programs constitute an indirect cost to the livestock sector and consumers."

*** 6.56 ***
United States. Economic Research Service. *Cotton: Background for 1985 Farm Legislation.* Agriculture Information Bulletin, no. 476. Washington, DC: U.S. Department of Agriculture, 1984.

U.S. cotton producers have frequently experienced excess production capacity, high stocks, and low product prices. Government programs since the 1930s have supported prices and attempted to adjust acreage and production to meet market needs, with varying degrees of success. Continuing issues for farm legislation include the appropriate kinds and levels of price and income supports and effective ways to manage production over the long run with improvement of production efficiency. The interrelationships between the cotton industry and other sectors of the U.S. economy and the need to compete in world markets are important considerations.

*** 6.57 ***
United States. Economic Research Service. *Dairy: Background for 1985 Farm Legislation.* Agriculture Information Bulletin, no. 474. Washington, DC: U.S. Department of Agriculture, 1984.

The U.S. dairy industry is primarily a domestic industry growing at about the same rate as the U.S. population. Government program costs have exceeded $1 billion a year since the 1979-80 marketing year and reached a record level of $2.6 billion in the 1982-83 marketing year. Continuing issues are the appropriate price support level and the proper formula or mechanism for attaining it.

*** 6.58 ***
United States. Economic Research Service. *Honey: Background for 1985 Farm Legislation.* Agriculture Information Bulletin, no. 465. Washington, DC: U.S. Department of Agriculture, 1984.

Since 1952 the government has supported the price of honey at between 60 and 90 percent of parity to provide market price stability to honey producers and encourage maintenance of sufficient bee populations for pollination. With honey support prices well above the average domestic wholesale price since 1981, domestic honey producers and packers have imported lower priced honey for domestic use and have sold domestically produced honey to the government.

*** 6.59 ***
United States. Economic Research Service. *Oats: Background for 1985 Farm Legislation.* Agriculture Information Bulletin, no. 473. Washington, DC: U.S. Department of Agriculture, 1984.

Oats acreage has trended downward since the 1950s. Most production is fed to livestock on farms where produced, and exports are relatively small. Price support loans have been available to oats producers since the late 1940s; however, deficiency and diversion payments were not made to them until 1983.

*** 6.60 ***
United States. Economic Research Service. *Peanuts: Background for 1985 Farm Legislation.* Agricultural Information Bulletin, no. 469. Washington, DC: U.S. Department of Agriculture, 1984.

The peanut program has led to surplus production and increasing government costs. Farm legislation in 1977 initiated a two-price poundage quota peanut program, which was continued under the 1981 farm act. The 1981 act suspended peanut acreage allotments and the poundage quota was decreased each year to eliminate excess peanuts supported at the higher of the two support prices. Continuing issues include whether to continue the current program or to include peanuts under a more general farm program.

*** 6.61 ***
United States. Economic Research Service. *Possible Economic Consequences of Reverting to Permanent Legislation or Eliminating Price and Income Supports.* Agricultural Economic Report, no. 526. Washington, DC: U.S. Department of Agriculture, 1985.

The Economic Research Service explores what would happen if current farm programs were not renewed. Reverting to permanent support programs, dating back in some cases to the 1930s, would raise price and income support levels significantly and greatly reduce the role of market forces in determining farm returns. Conversely, if all price and income supports had been eliminated in 1985, government intervention in the market would end, and supply and demand forces would determine farm returns. The report concludes that "Adopting either of these two outer bound policy alternatives would have significant and far-reaching impacts on farm operations, the agribusiness sector, the general economy, and ultimately the world market for farm products."

*** 6.62 ***
United States. Economic Research Service. *Rice: Background for 1985 Farm Legislation.* Agriculture Information Bulletin, no. 470. Washington, DC: U.S. Department of Agriculture, 1984.

Rice ranks ninth among major field crops in value of production and all production is irrigated, providing more stable yields than many other crops. Three classes of rice are produced--long, medium, and short grain--with long grain predominant. Domestic use and exports of rice rose sharply during the 1970s but declined in the 1980s. Between 1980 and 1983, rice stocks rose and prices fell, pushing rice program costs from less than a tenth to over nine-tenths of the value of U.S. rice production. Rice growers are adopting high-yielding, semi-dwarf varieties of long grain rice. Rising production capacity, weak export demand, and the type, level, and flexibility of income and price supports are continuing issues for farm legislation.

*** 6.63 ***
United States. Economic Research Service. *Sorghum: Background for 1985 Farm Legislation.* Agriculture Information Bulletin, no. 475. Washington, DC: U.S. Department of Agriculture, 1984.

Large sorghum harvests, greater corn and wheat feed use, and high foreign currency prices of sorghum helped raise U.S. sorghum stocks in the early 1980s. Government payments to sorghum producers climbed from one-seventh of total sorghum returns above cash expenses in 1980 to six-sevenths by 1983. Growth in demand for sorghum will likely come from exports, mainly determined by U.S. and foreign government policies, growth in foreign incomes and livestock output, and export credit availability. Continuing policy issues include the level and flexibility of price and income supports and policy effects on trade, the livestock sector, resources, consumers, and taxpayers. The report points out that "Corn and wheat policies usually have been major factors affecting the consequences of sorghum policy."

*** 6.64 ***
United States. Economic Research Service. *Soybeans: Background for 1985 Farm Legislation.* Agriculture Information Bulletin, no. 472. Washington, DC: U.S. Department of Agriculture, 1984.

U.S. soybean production has increased rapidly and now ranks second to corn in value. Much of the growth in demand has come from growth in world markets for beans, meal, and oil; but both soybean meal and oil face strong competition from other high-protein feed sources and highly substitutable fats and oils. Soybeans, a relatively new cash crop in the United States, have shifted to new production areas, being free of any acreage or production restrictions. Although soybeans are relatively free from direct government programs, production levels are affected by other farm programs. They are supported by a loan rate which, in most years, has been below prices received by farmers. Continuing issues center on determination of the support level and related trade issues.

*** 6.65 ***
United States. Economic Research Service. *Sugar: Background for 1985 Farm Legislation.* Agriculture Information Bulletin, no. 478. Washington, DC: U.S. Department of Agriculture, 1984.

There is a long history of government involvement in the sugar industry, first in 1789 to finance government operations, and then since 1894 to maintain a viable domestic sugar industry by protecting it during periods of low sugar prices. Sugar use has continued its decade-long downward trend as a result of replacement by less costly corn sweeteners and noncaloric sweeteners in many food and beverage products. Domestic consumption of sugar declined 20 percent between 1977 and 1983, domestic production increased, and world production remains at surplus levels as sugar is one of the major sources of revenue for developing countries.

*** 6.66 ***
United States. Economic Research Service. *Tobacco: Background for 1985 Farm Legislation.* Agriculture Information Bulletin, no. 468. Washington, DC: U.S. Department of Agriculture, 1984.

Tobacco is grown in 21 states on about 200,000 farms. Several types and kinds are grown, but flue-cured and burley account for more than 90 percent of total production. Because of high U.S. support prices, the strong dollar, and other factors, exports have declined and imports have risen during the last decade. Despite marketing quota cutbacks, loan stocks have risen sharply. Consumption of tobacco products has stabilized because of higher prices and health concerns. Growers and consumers, rather than taxpayers, bear most of the costs of operating the tobacco program.

*** 6.67 ***
United States. Economic Research Service. *Wheat: Background for 1985 Farm Legislation.* Agriculture Information Bulletin, no. 467. Washington, DC: U.S. Department of Agriculture, 1984.

Wheat growers have faced excess supplies so far in the 1980s and they face slow growth in exports for the near future. Even so, wheat production has remained large and yields are rising. None of the many farm policies employed in past years has succeeded in preventing periodic wheat surpluses or in reducing them without great expense.

The report concludes that "Issues for farm legislation include the level and flexibility of income and price supports, ways of increasing exports, and purpose and size of wheat reserves."

*** 6.68 ***

United States. Economic Research Service. *Wool and Mohair: Background for 1985 Farm Legislation*. Agriculture Information Bulletin, no. 466. Washington, DC: U.S. Department of Agriculture, 1984.

Wool and mohair have been declining, while sheep inventories have declined to a fifth of their World War II level and goat numbers are a third of their mid-1960s level. Economic recession and ample global supplies of wool and mohair in the early 1980s lowered prices of these fibers and reduced net returns of farmers while government payments to wool producers rose to record highs. The report points out that "Policymakers have had limited control over wool program costs given the formula-based government support price, the trend of declining textile market share, rising raw wool and wool textile imports, stagnant lamb and mutton consumption, and the dominance of Australia and New Zealand in the world wool market. Issues include whether to continue the program, and if so, the level and method of adjusting support prices."

*** 6.69 ***

Webb, Alan J., Jerry Sharples, Forrest Holland, and Philip L. Paarlberg. "World Agricultural Markets and United States Farm Policy." In *Agricultural-Food Policy Review: Commodity Program Perspectives*. Agricultural Economic Report, no. 530. Washington, DC: Economic Research Service, U.S. Department of Agriculture, 1985.

Agricultural exports are important to both the farm and the non-farm economies. The authors conclude that the United States "has the resources and technology to be competitive in global grain and oilseed markets, but farm policy will play an important role in determining the evolution of United States agriculture and its competitiveness on world markets. U.S. policies that support producer incomes are in conflict with strategies to expand exports. The policy choice is between support for today's producers and expanded markets for tomorrow's producers."

General Works

*** 6.70 ***

American Enterprise Institute for Public Policy Research. *Food and Agricultural Policy: With a Foreword by Don Paarlberg*. Washington, DC: American Enterprise Institute, 1977. ISBN 0-8447-2109-3.

The American Enterprise Institute organized this conference covering the full range of farm and food policy, with emphasis on price and production policy for the major farm commodities. As defined by Don Paarlberg, the real issue "is not the Democrats versus the Republicans, as once was the case. It is not the farmers versus the consumers, as some say it now is. The real issue is between the White House, which has concern for the entire economy, and the Congress, where the special interests are strongly entrenched."

*** 6.71 ***

Benedict, Murray R. *Can We Solve the Farm Problem? An Analysis of Federal Aid to Agriculture*. New York: The Twentieth Century Fund, 1955. ISBN 0-8447-2109-3.

In a comprehensive study designed for the general reader, the author has included a substantial summary chapter by a 12-member committee of economists on agricultural policy organized to make policy recommendations based on the study. The discussion generally concentrates on the principles in farm programs rather than the evaluation of individual programs. Benedict's conclusions are conservative in tone, strongly implying less government involvement, especially in farm commodity markets, a position with which the committee strongly agrees.

*** 6.72 ***

Bennett, Merrill K. *The World's Food: A Study of the Interrelations of World Population, National Diets, and Food Potential*. New York: Harper and Brothers, 1954. LC 53-11838.

Bennett provided the most comprehensive or complete study of the world food situation up to mid-century. He did not take either a generally alarmist or a sanguine view of world food prospects, but rather showed how the development of production efficiency and population growth tended to balance over the medium and long term. In a balanced evaluation, Bennett showed that world food problems are important and calls for judicious policies by both developed and developing countries.

*** 6.73 ***

Black, John D., and Maxine E. Kiefer. *Future Food and Agricultural Policy, a Program for the Next Ten Years*. New York: McGraw-Hill Book Company, 1948. LC 48-10583.

In this work, Black and Kiefer have written what is arguably their most comprehensive book. It presents a very broad and detailed program, tending to deemphasize production controls in favor of a more balanced and broader perspective, emphasizing food abundance, market growth, and expanding world trade.

*** 6.74 ***
Bogue, Donald J. *Principles of Demography*. New York: John Wiley and Sons, 1969. ISBN 0471087620F.

A distinguished demographer has written this scholarly, broad-gauged book that takes a rather conservative view of population growth. Bogue shows that population growth is not uncontrolled, as some more alarmist studies were portraying at the time, but tends to conform to human values and beliefs. Growth can be regulated by personal decisions and publicly supported programs.

*** 6.75 ***
Breimyer, Harold F. *Farm Policy: 13 Essays*. Ames: Iowa State University Press, 1977. ISBN 0813806453.

In the words of the author, these essays are "discursive and neither closely interconnected nor free of overlap. They are intended to prick, stimulate, stir imaginations, and above all to teach. Although opinions are prevalent, the book is not a tract of advocacy."

*** 6.76 ***
Cochrane, Willard W. *Farm Prices: Myth and Reality*. Minneapolis: University of Minnesota Press, 1958. LC 58-7556.

Cochrane argued the case for strong farm production and marketing controls as a means of raising and stabilizing prices of farm commodities. His general philosophy became the cornerstone of the farm policy supported by the Kennedy Administration in the early 1960s, when Cochrane served as the chief policy economist in the Department of Agriculture.

*** 6.77 ***
Cochrane, Willard W., and Mary E. Ryan. *American Farm Policy, 1948-73*. Minneapolis: University of Minnesota Press, 1976.

The authors present a historical account of farm policy, with an emphasis on commodity policy, and generally express a supportive view of government programs in supply management. Although its views on the benefits of price supports may arouse controversy, the book is factually correct and theoretically interesting.

*** 6.78 ***
Curry, Charles E., and Wm. Patrick Nichols. *Agriculture, Stability, and Growth: Toward a Cooperative Approach*. Port Washington, NY: Associated Faculty Press, 1984. ISBN 0-8046-9383-8.

In this volume, the authors prescribe a group of general policy orientations for agricultural stability and growth. Martin Abel and Lynn Daft, in the summary chapter, create a framework for a "best case" and a "worst case" scenario for the circumstances surrounding agriculture and the policymaking process.

*** 6.79 ***
Griswold, A. Whitney. *Farming and Democracy*. New York: Harcourt, Brace and Company, 1948. LC 48-6511.

The author sheds light on the thesis that farming is crucial as a positive force for democratic government. Griswold, who generally rejects the thesis, presents a number of examples from the United States and other countries in support of this rejection. His view is strengthened by the continuing decline of the farm population in the United States.

*** 6.80 ***
Guither, Harold D. *The Food Lobbyists: Behind the Scenes of Food and Agri-Politics*. Lexington, MA: Lexington Books, D. C. Heath and Company, 1980. LC 79-6734.

The author states his purpose as: "to identify and briefly describe the many organizations and groups that have had vital interests and concerns about federal government decisions in some phase of agriculture, food production and distribution." It is a reliable directory and analysis of the organized forces that seek to influence the course of farm and agribusiness legislation. It describes the origin and development of these organizations, their general political position on many issues, some of their successes and failures, and their influences on policy.

*** 6.81 ***
Hadwiger, Don F., and Ross B. Talbot, eds. *Food Policy and Farm Programs: Proceedings of the Academy of Political Science*. New York: The Academy of Political Science, 1982. LC 82-70800.

Like its predecessors in a series that started in 1910, this 135th issue of the proceedings deals with major "political, economic and social issues and to provide a link between the academic community and the world of public affairs." The volume attempts to deal with "the substance of farm policy"; a leading question being, "How efficiently do present policies serve us?" The academy serves as a public forum, but, as an organization, makes no recommendations on political questions.

*** 6.82 ***
Halcrow, Harold G. *Agricultural Policy Analysis*. New York: McGraw-Hill Book Company, 1984. ISBN 0-07-025562-8.

Halcrow has designed this book as a comprehensive text for the first general course in agricultural policy or agricultural and food policy. It deals with the public measures that can be taken to improve the competitive structure, operation, and performance of American agriculture in two broad sets of markets: those for products and those for inputs. The approach is from the general to the specific, dealing first in part one with the concepts of values and goals, the setting for policy, and the general methodology that is used. Part two develops the foundations for policy in the product markets and the alternative choices for the three economies of crops, livestock, and marketing. Part three covers policy in the input markets--the financial, natural resource, agribusiness input, and human resource markets.

*** 6.83 ***

Halcrow, Harold G. *Food Policy for America.* New York: McGraw-Hill Book Company, 1977. ISBN 0-07-025550-4.

Halcrow offers a general study of farm, food, and agricultural policy in the United States, covering the history of this policy, the major legislative acts and programs, and the methods of economic analysis used in appraisal of the policies and programs. It is designed for general readers interested in these topics, and for university students studying policy.

*** 6.84 ***

Heady, Earl O. *Agricultural Policy under Economic Development.* Ames: Iowa State University Press, 1962. LC 62-9124.

Heady intends this book mainly for students, professional economists, and policy administrators. It combines a fair amount of theory and analytical treatment with the more descriptive and literary analysis of agricultural structure and policy. While it does include some equations and graphs and a frequent focus on technical terminology, it also has a major content in purely literary manner. In the words of the author, "There are very large 'sketches' which do not include technical terminology and can be read independently."

*** 6.85 ***

Heady, Earl O., Leo V. Mayer, and Howard C. Madsen. *Future Farm Programs, Comparative Costs and Consequences.* Ames: Iowa State University Press, 1972. ISBN 0813806755.

In this work, the authors deal with the problems of farm programs in the commercial sector of U.S. agriculture. They state that "It provides an in-depth analysis of alternative land retirement or diversion programs which use price support, direct payments, and other methods for supply control or restraint. Hence it provides analyses of a restricted set of agricultural policy alternatives since it relates mainly to the use of land retirement, withdrawal, or conversion as basic policy mechanisms."

*** 6.86 ***

Knutson, Ronald D., J.B. Penn, and William T. Boehm. *Agricultural and Food Policy.* Englewood Cliffs, NJ: Prentice-Hall, 1983. ISBN 013089111.

The authors have designed this work as a textbook for undergraduate courses in agricultural and food policy. Their main thesis is that agricultural and food policy encompasses four problem areas: farm prices and incomes; world food and trade; consumer food prices and availability; and natural resource considerations. In their words, "It is all too easy to fall into the trap of treating each of the four agricultural and food problem areas separately....Students of policy must continuously be looking at causes and effects that extend beyond the immediate problem or policy alternatives."

*** 6.87 ***

Paarlberg, Don. *Farm and Food Policy.* Lincoln: University of Nebraska Press, 1980. ISBN 0803236565.

Paarlberg presents a broad, accurate delineation of U.S. farm and food policy with emphasis on the nation's experience with commodity price support programs. Generally conservative in respect to policy, Paarlberg concludes that the nation will be best served by a more moderate policy emphasizing a primary reliance on free markets rather than on production controls and large government expenditures.

*** 6.88 ***

Rodefeld, Richard D., Jan Flora, Donald Voth, Isao Fujimoto, and Jim Converse. *Change in Rural America: Causes, Consequences, and Alternatives.* St. Louis: C.V. Mosby Company, 1978. ISBN 0-8016-4145-4.

The authors' main objective in this volume is to provide the reader with a better understanding of twentieth century changes in the rural sector of the United States. They do this by focusing on initial changes that have occurred in several areas: agricultural technology; farm organizational and occupational structure; transportation; and the rural economic base. They cite articles that discuss the topics in more detail, argue positions divergent to those presented, and discuss relevant topics not addressed in their volume. Generally, references that can be easily and inexpensively obtained are not included in the text, although many of these have been included as references.

*** 6.89 ***

Rosenblum, John W., ed. *Agriculture in the Twenty-first Century.* New York: John Wiley and Sons, 1983. ISBN 0-471-88538-X.

The substance of a two-day symposium on "Agriculture in the Twenty-First Century" is presented in this volume. The symoposium was attended by academic, government, and business leaders from around the world. It explored the latest available knowledge, applicable to the United States and other countries, on a wide range of topics in food, farm, and resource policy.

*** 6.90 ***

Ruttan, Vernon W., Arley D. Waldo, and James P. Houck, eds. *Agricultural Policy in an Affluent Society: An Introduction to a Current Issue in Policy.* New York: W.W. Norton and Company, 1969. LC 69-12587.

The editors have compiled a collection of 26 articles, mostly by well-known agricultural economists, which is still relevant for students of farm policy. It succinctly summarizes the major elements in a still unresolved issue in farm policy: how best to adjust the high technology, modern agriculture of the United States to the affluence of the national market and the larger more mixed economy of the world market. A number of suggestions are made by the authors for adjusting the farm and rural economy to modern conditions.

*** 6.91 ***

Schultz, Theodore W. *Agriculture in an Unstable Economy.* New York: McGraw-Hill Book Company, 1945.

In what was the first and arguably both the most controversial and widely read book in the post-war era dealing specifically with general farm policy, Schultz presented a unique program for farm income support based on prices for major farm commodities set ahead at least one production period, which would be called "forward prices." No action would be taken by government to make these prices good in markets. But, providing unemployment rose above a certain level, which was suggested by Schultz to be four percent of the national labor force, "compensatory payments" would be paid to farmers to cover part or all of the difference between the forward prices and market prices, assuming the latter were lower. These proposals initiated widespread discussion, but were not adopted by Congress.

*** 6.92 ***

Schultz, Theodore W., ed. *Distortions of Agricultural Incentives.* Bloomington: Indiana University Press, 1978. ISBN 0-253-31806-8.

A three-day workshop on resources, incentives, and agriculture was sponsored by the American Society of Arts and Sciences in 1977. Invited papers and comments were prepared by people with expertise in agricultural development, both national and international. The papers highlight many of the distortions on agricultural production and trade imposed by government interventions, such as production and marketing subsidies and control, tariffs and import quotas, export subsidies that are most often used by rich countries, and export taxes sometimes imposed mainly by poor countries.

*** 6.93 ***

Schultz, Theodore W. *Transforming Traditional Agriculture.* New Haven: Yale University Press, 1964.

In this relatively small contribution, Schultz wrote what is arguably his most noted book. Based on a series of lectures at Yale University, it presents the thesis that farmers must have appropriate economic incentives to transform their farms from traditional labor-intensive types of enterprises to modern capital-intensive and larger scale firms. The evolution of this proposition has had a significant impact on farm policy and development. The thesis is important in a review of policy for times ahead.

*** 6.94 ***

Tweeten, Luther C. *Foundation of Farm Policy.* 2d ed., rev. Lincoln: University of Nebraska Press, 1979. ISBN 0-8032-0972-X.

Tweeten states his purpose as: "to provide the foundation needed to understand, interpret, and analyze farm policy," to be "used as a text, as a reference, or for selective reading." It is a broad, comprehensive book dealing with policies that affect groups and not just individuals, concentrating on the ways that farmers influence and are influenced by government policies.

CHAPTER 7

TRANSFORMING FARM POLICY FOR TIMES AHEAD

The financial crisis of the American farm was created by the mix of past and current policy. It will intensify until something fundamental is changed. Major changes should include reduction of government costs, improving quality and quantity of exports, and refinancing and restructuring of the farm debt.

"The U.S. farm policy of the future must be geared to compete for buyers who have more alternative sources of supply than ever--their own agriculture, competing agricultures all over the globe, and more synthetics and substitutes....the price supports, land diversion, and storage programs that have dominated U.S. farm policy for the past 50 years work against the U.S. farmer in a world of high technology and rising productivity."

Dennis Avery[1]

"Perhaps the most difficult problem is how we move toward a longer-term rational, comprehensive, coherent and just policy when there is an immediate crisis in much of agriculture."

Alex McCalla[2]

The financial crisis of the American farm did not just happen. It has been created by the mix of past and current policy. It will intensify year by year until something fundamental is changed. What can be changed, and how it can be changed is the main topic of this chapter.

Commodity Program Alternatives

By the mid 1990s, under current commodity programs, as much as 85 to 90 percent of the market output of American farms will be produced on 300,000 or fewer large farms. The small and medium family farms, large enough to absorb the full-time of a family but not large enough to make the most efficient use of modern technology, will largely be gone. Their rates of exodus will be speeded up by the constraints on production and the distribution of program benefits under production control programs. These favor the larger farms.

Many smaller residential farms with good off-farm sources of income will survive. Commodities produced on these farms and on the few remaining small and medium size family farms will supply 10 to 15 percent of the farm commodity market.

If present commodity programs are continued, however, the American farm will not reach its full production-efficiency potential. Costs will be increased by resource restrictions on output, including acreage bases and allotments and conservation reserve programs. These encourage farmers to compete more aggressively for crop land to get a larger program base, which pushes up production costs. The higher costs then produce an excuse or rationale for continuing and intensifying the programs. Marketing agreements, which regulate and limit the sale of some crops (such as some California oranges, lemons, dates, and some fruits, vegetables, and nuts), also help to raise costs by protecting producers and weakening interfarm competition.

Although the production control programs are mainly limited to wheat and feed grains, cotton, rice, peanuts, and tobacco, the effect is felt all through the farm economy. Producers of these commodities generally have not been more prosperous and will not have higher rates of return on their investments than producers of commodities not directly under government control, including producers of beef, pork, and poultry, and most fruits and vegetables. The higher costs attributed to controls tend to wipe out the effects of higher commodity prices, and profits tend to equalize between or among sectors. In some cases, higher prices for some commodities, such as feed grains, are translated into higher costs for other producers, such as beef, dairy, pork, and poultry producers. This is the reason that the American National Cattlemen's Association and the National Pork Producers Council, the main representatives for beef and pork producers, have wanted to phase out commodity programs.

The dairy price support program, designed to help hard-pressed dairy producers, has been widely criticized as a very expensive program, costing $4 to $5 billion a year and failing to stabilize net incomes of dairy farmers. In 1983, because of large government purchases required to support prices, running up to more than 10 percent of total dairy production, Congress authorized a voluntary milk reduction program. Later the 1985 Food Security Act included a Dairy Termination Program (DTP), as a means of reducing dairy production by total herd liquidation. This was approved for only 18 months in 1986-87.

Congress continued price supports at levels sufficiently high, however, to encourage more production. Although the DTP had an immediate effect in helping some farmers to discontinue dairy production, producers not in the program could continue to expand output. The total number of milk cows on farms dropped two to three percent, while favorable milk-feed price ratios stimulated increased concentrate feeding and higher output per cow. By 1986, milk output per cow was more than 2 percent above 1985, and about 6 percent over 1983.[3] It appears that as long as price supports are high enough to encourage more dairy production, the effects will be high government costs, more attrition among family-farm-type dairy producers, higher prices for consumers of dairy products, and unwanted surpluses.

Under the 1985 bill, the federal government is committed to a subsidy of $25 billion a year, possibly much more in some years, to control production and support farm incomes. Nevertheless this huge subsidy, more than an average of $12,000 per farm for the nation's 2.2 million farms, has not been large enough to offset the losses in net farm income due to lost markets, idled land, and the cost-increasing effects of price supports and other programs. At the same time, the U.S. trade deficit is increased by the effects of price supports on exports of farm commodities. In addition, perhaps irrationally, the reactions of foreign governments have been to tighten their restraints on trade, as producers in these countries feel themselves to be threatened by the U.S. programs and subsidies. As long as these continue, the United States may witness a further tightening of export markets. Resistance to freer markets by producers in other countries will increase. History tells us that government intervention invites, or forces, more intervention. The United States, as the largest world trader in farm commodities and most other goods, has the option of continuing the current costly policy, or transforming its commodity programs for the times ahead.

Two major and quite different proposals were introduced in the 99th Congress as substitutes for the 1985 farm bill. The first was introduced by Senator Tom Harkin (Democrat-Iowa), entitled "Save the Family Farm Act." It would raise price supports much higher than provided in the 1985 bill, tighten production and marketing controls, while liberally supplying government-owned commodities at world prices to maintain the U.S. share of farm export markets.

The second was introduced by Senator Rudy Boschwitz (Republican-Minnesota) and Senator Boren (Democrat-Oklahoma), entitled the "Family Farm Protection and Full Production Act of 1985." It would do away with acreage allotments and most marketing controls, base farm subsidies that decline year-by-year on records of past production, strengthen world food-aid programs, and move toward free trade in farm commodities.

Neither of the two bills was passed, although the Boschwitz-Boren bill received 42 votes in the Senate, only 9 votes short of passage. The Harkin bill failed by a much larger margin, but in 1987 it was introduced in the Senate again by Harkin and in the House by Representative Richard Gephardt (Democrat-Missouri).

The provisions of the Harkin bill were broad, covering price supports, supply controls, dairy policy, international trade, food assistance, and farm finance. As introduced in the 99th Congress, price supports would be set in the first year of operation in 1988 at 70 percent of parity, as compared with a parity ratio, which measures the price level of all farm commodities relative to non-farm prices, hovering around 50 percent in the mid 1980s. Then the bill proposed to raise the support level by one point each year through the year 2000. For instance, there would be 70 percent of parity in 1988, 71 percent in 1989, 72 percent in 1990, and so on. By 1990, supports would be 15 to 25 percent above the *target price levels* in the 1985 farm bill.

The strong production incentives provided by these support levels would be curbed by mandatory set-aside, acreage allotments, marketing quotas, and marketing certificates. Mandatory supply controls would be imposed on approval of *50 percent* of producers voting in commodity-specific referenda (historically a two-thirds majority has been required). The maximum set-aside would be 35 percent of a producer's base on the largest farms, and the required set-aside would be reduced by 1 percent through 10 farm sizes, down to 25 percent on the smallest farms. All producers (including farmer-feeders), however, might be required to set aside up to 35 percent of their base when stocks are excessive.

A national marketing quota was provided, based on domestic and export demand, food-aid programs, carryover levels, and reserve requirements. Producers could only market commodities for which they held certificates based on their permitted acreage (non-set-aside land), and farm program yield, or county yield, if higher. But farmers could be allowed to feed their over-certificate production on the farm where grown, presumably a boon to grain-livestock producers. Also, production above the amounts covered by certificates could be stored on the farm or sold to the CCC at a maximum of 50 percent of the support price.

The Harkin bill would involve a two-price plan, which separates the high prices in the domestic market from the lower prices in the foreign market, an idea that has been debated in farm-policy circles since the 1920s. The effectiveness of the plan, if it were to become law, would depend on three crucial factors: (1) the effectiveness of production and marketing controls in offsetting the production-stimulating and consumption-decreasing effects of higher domestic prices; (2) the commitment of the United States to export farm commodities at world prices either through use of government stocks or direct subsidy; and (3) the ability of the government to prevent imports of competitive farm-food commodities and fibers.

The ability of the United States to prevent imports of commodities that are substitutes for price supported commodities may be questioned. Much higher tariffs and, in some cases, strict quotas would be required. The political costs in stifling world trade would be great.

The Boschwitz-Boren bill contained provisions to base any farm subsidy on records of past production, while discontinuing price supports and dropping production controls. It contained specific targets for exports and suspension of allotments, and set a schedule for loan rates and transitional payments to 1993 for wheat, feed grains, cotton, and rice. Provisions for a conservation reserve and a programmed grain reserve were included. The bill would repeal the 1985 dairy program, replacing it with lower price supports and a system of payments to producers through 1990.

The most unique and critical feature of the Boschwitz-Boren bill was the proposal to base payments to grain and cotton producers on the acreage base for 1984 and 1985, and an established yield, as recorded for 1985, or 1984-85. Payments would have been scaled down from 100 percent for 1986, to no more than 92 percent for 1987, 80 percent for 1988, 65 percent for 1989, and 50 percent for 1990. Under the dairy program, payment limitations per producer would have been $20,000 for fiscal 1986, $18,400 for 1987, $16,000 for 1988, $13,000 for 1989, and $10,000 for 1990. Farmers would receive a definite payment regardless of what was seeded in any year, or whether market prices rose or fell.

The bill provided that the total amount of payments that a producer would be entitled to receive for wheat, feed grains, cotton, rice, and dairy could not exceed $63,000 for the 1986 crop, $57,960 for 1987, $50,400 for 1988, $40,950 for 1989, and $31,500 for 1990. Payments in excess of $20,000 but not more than $60,000 would be reduced by 50 percent, in excess of $60,000 but not more than $120,000 by 75 percent, in excess of $120,000 but not more than $200,000 by 90 percent.

The three major goals of (1) maintaining farm income, (2) making U.S. commodities price competitive in world markets, and (3) limiting federal budget costs could be achieved over time by such a program. Presumably, prices on the major export commodities would drop to near world market prices, as was nearly achieved under the 1985 farm bill. The maintenance of farm income would depend on the adequacy of the subsidy, and the income effects of production adjustments made by farmers. Since the subsidy would drop year-by-year, such adjustments would be crucial. Farmers would be free to plant what they wanted instead of being locked into a fixed historic base or being required to plant so much of a crop to protect their base. Some marginal land would go out of production or shift to another use. Farmers would produce the commodities having a comparative advantage, which is most efficient and profitable.

It is difficult to fully compare the Harkin and Boschwitz-Boren bills. This is especially true in respect to the quantitative results that might ensue. To be successful, a program must do efficiently what it is designed to do and it must win public support. Efficiency is usually measured in terms of benefits and costs. Public support or acceptance depends on the public's perceptions of these measures as compared with prevailing values and beliefs about what should or should not be done.

Because of high price supports, the Harkin bill would give an immediate boost to farm income with windfall-capital gains to owners of farm land and other real estate. But after a short period of adjustment, the gains in net income of farm operators would be constrained by higher costs for land and other inputs for which farmers compete. The competitive nature of farming assures that net incomes of farm operators do not stay above a normal equilibrium level for very long.

The Boschwitz-Boren bill would not provide any short-term income gain to farmers. But, by projecting a specific subsidy over a minimum of five years, it would provide an income cushion as farmers adjust to market prices. Whether net farm income would increase or decrease would depend on world demand and trade, management of storage stocks, as well as on trends in farm production. Production trends in livestock, fruits, and vegetables depend heavily on domestic demand, rather than farm programs. Production of wheat, feed grains, rice, and cotton, the major commodities that have been under government programs, would be more sensitive to market forces.

Net farm income could drop under the Boschwitz-Boren bill, at least for a short period, but any drop in market receipts could be cushioned by direct subsidy payments. According to the schedule of payments proposed in the bill, the small and medium family farms would be helped the most.

Neither bill would necessarily "save the family farm," however. The proposals for set-aside in the Harkin bill would progressively favor the smaller family farms, but the acreage restriction and marketing quotas would further limit the use of farm resources, especially on the smaller farms. Increased competition for land resulting from these restrictions would push up land prices and costs.

Although the subsidy schedule in the Boschwitz-Boren bill favors smaller farms, the phasing down of subsidy payments would leave all farms more open to purely competitive forces. The smaller farms would not be penalized but would continue to decline in number as advancing technology favors larger farms.

In terms of consumer welfare, budget costs, and trade, the Boschwitz-Boren bill is preferable to the Harkin bill. Under the latter, consumers would pay much more for food and other farm products. Budget costs to implement acreage allotments and exports would be much higher, running upward from $50 billion a year under 1988 conditions and increasing each year thereafter.

Finally, the performance of the U.S. government in foreign markets would be improved by being able to offer commodities at price-competitive levels, as both bills provide. If the United States is to avoid charges of "dumping," or selling below "cost-of-production," which tend to be countered by restrictions on trade, however, clearly the comparison favors the Boschwitz-Boren proposals.

In the event that price supports were not eliminated but merely phased down under the Boschwitz-Boren bill, some land retirement might be required to prevent accumulating storage stocks larger than were wanted. Moreover, it is prudent under any bill to continue land conservation, emphasizing retiring the most highly erodible cropland from crop production and regulating grazing on public lands.

World and Domestic Market Perspectives

Population growth and income in comparison with farm productivity and growth in output will largely determine the future world markets for farm commodities. During this generation and the next, population growth will greatly expand the world market potential, while changes in income and farm productivity will determine whether the potential is realized or not. Because of the high income in the United States and anticipated slow population growth, future markets for the American farm will be determined more by what happens around the world than by what happens in the domestic economy.

Because birth rates are continually changing, and continuing to drop in important world regions while longevity is increasing, the total world population to be expected at any future date is always imperfectly predictable. The further the projections are extended, the more uncertainty there is. For instance, from a total world population of 5 billion people in 1985, world population should reach 6 billion by 1995 or 1996. It took just 10 years to increase from 4 to 5 billion. By 2020, the total could be anywhere from about 7.8 to 10 billion.[4]

With continuing research and development, advances in technology will be capable of maintaining and gradually increasing per capita food production up to 1995 (as the projections in Chapter 2 suggest). This will require 1985 world grain yields to increase 20 percent by 1995 with corresponding increases in livestock numbers *or* productivity. A world population of 7.8 to 10 billion people in 2020 will require more than trend increases in research and development if world per capita food production is to be increased significantly. It will require a 56 to 70 percent increase in grain yields, with corresponding increases in output of livestock. The area on which grain is grown will expand slowly, as it has over the last 25 to 50 years. The carrying capacity of most of the world's grasslands will have to be improved significantly, but gains equivalent to those in grain yields appear to be out of reach. Except for the growth of aquaculture, or "fish farms," the world's fish catch is at, or nearly at, its maximum sustainable yield.

To compare the growth of markets in developed and developing countries, food needs must be distinguished from food demand. Population determines food needs, but the rate at which needs are translated into demands depends on the availability of income to buy food. Most of the people in developed countries have sufficient income to purchase well-balanced diets, and welfare programs exist or can be improved to help most of those in poverty. Over the next generation, most developed countries, with the possible major exception of Japan, will increase food imports very little and some not at all because of either slow population growth or their increasing farm productivity, or both. The developing countries generally face a different prospect.

There is a close race in most developing countries between population growth and food production. Of the 5 billion people in 1985, some 3.7 billion were in developing countries. The United Nations Population Division, which has prepared long-term "low," "medium," and "high" projections, shows in its low projection that this 3.7 billion will at least double in 45 years to 7.4 billion by 2030, and could rise to 8.2 billion or 9.1 billion people.

From 1950 to 1985, there was an unprecedented increase in world grain production from 620 million metric tons to 1.66 billion tons. The average yield per hectare (about 2.47 acres) rose from 1.1 tons to 2.6 tons, but yield increases were highly concentrated on less than one third of the 425 million hectares planted to grain in 1985, with countries in Asia making rapid strides and most in Africa increasing very little.

The implications of these projections for the American farm are far-reaching, indeed. Long-term growth in food needs is clearly targeted, if not exclusively targeted, in the developing countries. The major and most stable markets for U.S. farm commodity exports have been Western Europe and Japan. But since its recovery from World War II, Western Europe has been moving toward food self-sufficiency, a condition now practically achieved. By the early 1990s, Western Europe will be a consistent net food exporter. Japan, which has been the largest single importer of U.S. farm products, will increase its food imports slowly, at least up to 1995. Japan's farm production is heavily subsidized and, due to a combination of national policies, its rate of population growth has been slowed remarkably.

Currently, Japan produces about 11 million metric tons of rice, which is of course its most important staple food commodity. Rice production uses a lot of the best land in Japan. It has been pointed out that, if protection to rice growers were reduced, some of this land could be used to lessen the overcrowding caused by urban growth.

Whether the developing countries become strong markets for farm exports from the United States and other exporting countries depends on their progress in developing more efficient and diversified economies. In the period between 1961 and 1980, developing countries' cereal production grew at an annual rate of 2.9 percent a year, while consumption grew at the considerably faster rate of 3.2 percent a year. As a result, net annual cereal imports of the developing countries increased more than fourfold in the period, from about 15 million to 64 million tons. From the late 1960s to the late 1970s, the 29 developing countries with the fastest growth rates in basic staple food production increased their imports of basic food staples by 360 percent. A standard projection of such trends to 2000 suggests an increase of 40 million tons over 1980 in imports of staple food crops by all developing countries. The 1984 actual net imports were on the projected trend line. Only a few developing countries were net exporters of food, with Argentina and Thailand accounting for 68 percent of total developing-country cereal exports in 1981-83.[5]

Developing countries have been increasing their imports of livestock products at a rapid pace, although the total values are far below their imports of cereals and other crop staples. Because the demand for livestock products increases strongly with per capita increases in income, more livestock products will be consumed in the developing countries.

Livestock production is labor intensive, especially in these countries; therefore, it is logical for developing countries to produce most of the livestock products they consume. However, if they displace with their own production most of the increase in livestock products that otherwise might be imported, they will increase their imports of cereals and other staples for use as livestock feed. By 2000, this trend could add at least another 40 million tons of feed grains to their total imports.

These projections depend on the continued broad and diversified growth of the developing countries. Assistance by the United States and other developed countries in helping to create these necessary conditions for growth should be a matter of high priority.

World economic development and free world trade are the hope for the American farm in the long view. To help bring this about should be the major policy interest of American farm leaders and farm organizations. The urgency of this policy can be evaluated further by considering factors in projecting the growth of the domestic market for farm commodities.

Domestic Market for Food

The domestic market for various foods and other farm commodities depends on consumers' income and its distribution; population growth and age distribution; household size; labor force participation and population mobility; and health and nutrition information. Although each of these factors is important, they are not of equal weight.

1. *Consumer's income and its distribution.* As per capita income increases, a smaller and smaller percentage of income is spent on food, and the consumer's responsiveness to changes in food prices decreases. For example, upper income consumers in the U.S. spend about 11 percent of their income on food while the lower income households spend 40 percent, or more. Middle and upper income people buy the same quantity of food, although not necessarily the same quality, regardless of modest price changes.

In the last 20 years in the United States, real per capita income has risen about 2.5 percent a year while the average calorie intake has dropped 0.5 percent a year. Although there is some disagreement about the effect of rising real incomes on the demands for specific foods, there is general agreement that rising incomes result in increasing demands for a greater variety of foods and more convenience in their delivery. More variety is found in increasing consumption of meat substitutes, various cheeses, nuts, fresh or frozen fruits and vegetables and their juices. More convenience means expansion of food processing and packaging services, growth of restaurants and eating out, without increasing the demand for more food, or more food calories.[6]

Although it is very difficult to distinquish the effects of rising real incomes from the effects of greater knowledge about nutrition and health, recent experience suggests that further increases in real income will not increase per capita demand except among the population at the lower end of the income scale. Although raising the real incomes of the 15 to 20 percent of the population who have had the least advantages would increase per capita demands for this group, the effects on average per capita demand for food would be relatively small.

2. *Population growth and age distribution.* Because of high real incomes in the United States, most of the increase in demand for food will depend on the long-term growth in population. Over the past 30 years population growth has averaged about 1.3 percent a year. Based on current immigration policy and demographic trends, population is expected to grow much slower in the next 30 years, perhaps only half as much.

Consequently, the rate of growth in domestic demand for all food will be lower than it has been. Even an upgrading in the quality of the diet, such as an increase in consumption of animal products, will not upset this trend, especially in respect to demand for feed grains. That is, increases in the efficiency with which animals use feed will offset increases in the consumption of animal products. Hence, even with an upgrading of the diet the increase in domestic demand for farm commodities due to population growth will be less than 1 percent a year.[7]

The changing age distribution of the U.S. population is not likely to upset this basic trend. As the population grows, the numbers and percent of elderly persons continues to increase. The elderly segment of the population is increasingly healthy, affluent, and predominately female. Elderly persons usually spend less per person for food since daily caloric requirements decline with age. For instance, the recommended average daily allowance of calories for women drops from 2,100 at age 19 to 1,650 at age 65. Also, elderly persons typically buy more poultry, fruits, vegetables, baking products, and cereals; and they spend less on red meats, milk, and soft drinks. They have tended to spend a smaller portion of their food dollar on eating out, but with increasing affluence this may change.

Nevertheless, perhaps the only real increase in per capita expenditure that will affect farmers directly may come from closing the income gap on the small minority of the population that has not had adequate income for a good diet.

3. *Household size.* The major trends in house-

hold size have accentuated rather than offset the effects of changing age structure. The average household size in the United States has decreased from 3.8 persons in 1940 to 2.7 persons in 1985, and is projected to decrease to 2.4 persons by the year 2000. Nearly one-fourth of all households have only one member while 55 percent have two or fewer members. These smaller households spend 44 percent more per person for food and a much higher percentage of this for convenience, including meals away from home. For instance, single people spend on the average half of their food dollar eating out, with the more affluent averaging more. The smaller households spend more on poultry, fruits and vegetables, except potatoes, fish, cheese, bakery products, except bread and cereal, and soft drinks. They spend relatively less on beef, pork, fresh dairy products, eggs, sugars, sweets, and processed vegetables.

The decreases in average household size are due to lower birth rates, higher divorce rates, marrying later or not at all, and increasing longevity. The food trends associated with this do result in higher expenditures on food and drink. But most of the income is absorbed in more convenience and service, rather than on more demand per capita.[8]

4. *Labor force participation and population mobility.* Over the last four or five decades, increased participation of women in the labor force has increased real incomes and decreased leisure time. Also increased is the demand for variety and convenience in food, but not necessarily the total demand for food. In 1986, almost 70 percent of women aged 25 to 44 were in the labor force, and 73 percent of them worked full-time as compared with 86 percent of working men. Households with working wives had median weekly earnings 51 percent higher than households where only the husband worked. In 1984, one-fifth of the working wives had higher incomes than their husbands. If a high percentage of women in the 25 to 44 age group continue to work as they grow older, then this will tend to increase demands for variety and convenience without increasing the total demand for food.

Similarly, demographic shifts in the population tend to increase the demand for variety and convenience without increasing over-all demand. Increased immigration and a growing proportion of nonwhites in the population tends to increase variety. For instance, the growing popularity of Mexican, Oriental, Italian, and other ethnic foods suggests an increasing preference for variety without increasing the overall per capita demand.

5. *Health and nutrition information.* In recent years food science research and education have brought increasing awareness of some of the important relationships between diets, health, and longevity; and there is evidence that eating habits of some of the domestic population are changing sufficiently to produce a significant trend. The variety of foods eaten is increasing and concerns about being overweight apparently have influenced the types and quantities of food eaten.

Per capita consumption of foods high in calories has declined while the pounds of food consumed per capita has increased. For instance, national food consumption surveys show that the per capita calorie intake decreased from 2,036 in 1965 to 1,826 in 1978, while the annual consumption of food in pounds per capita increased from 1,381 in 1960-63 to 1,401 in 1980-83.[9, 10]

6. *Conclusion.* Since all consumers do not change their eating habits at the same time, the changes that have been observed can continue and accentuate for several more years, even decades. Trends are well established toward still more variety and convenience, more dietary fiber and lean protein, more fruits and vegetables; but fewer calories, less animal fat, and very slow growth in the over-all size of the domestic market. These trends will put more downward pressure on prices of traditional foods in excess supply, including grains, red meat, and dairy products.

The farm economy cannot be extricated from depression by cutting back on production and shutting itself off from world markets. Hope for the future lies in shifting toward those demands that are increasing in the domestic market and toward a stronger competitive posture in world markets.

Improving the Quality of Grain Exports

Importers have complained for many years about the quality of U.S. grain exports. In the 1980s, as the world market shifted in favor of buyers, complaints about the quality of U.S. exports multiplied. Research on the problem, which had begun at the University of Illinois in the early 1960s, had grown to encompass regional and national support. An important milestone was reached in the Grain Quality Improvement Act of 1986 (Public Law 99-641).[11]

Grain that graded satisfactory in U.S. elevators was often arriving in overseas ports out of condition. Some grain shipments were found to be moldy or musty, with some sprouting of kernels, and an excessive amount of broken kernels and foreign material, such as weed seeds, dust, and dirt.

The reasons for the out-of-condition grain were found to be inherent in the U.S. grain-grading standards, the methods of loading and unloading practiced by grain shippers, and the deteriorating effects of long ocean voyages.[12] Moisture levels, which were satisfactory for grain that was to be stored in aerated bins in the United States, and subjected to a minimum of handling, were found to be too high for grain that was to be handled three or four times after it left the farm, and placed in a ship's hold without aeration. Grain that passed inspection in the United States, but had been dried at too high a temperature, had a tendency to crack when handled by the high-speed pneumatic loaders and unloaders used by ocean-shipping grain firms. Broken kernels at standard moisture levels were susceptible to heating and spoilage. The worst problems tended to center around corn but all grain was involved to some extent.

The Grain Quality Improvement Act of 1986, effective November 10, 1986, established new standards to promote marketing of grain of high quality and to ensure the quality of grain marketed in or exported from the United States. The act required the secretary of agriculture to develop new standards to define dockage and foreign material and establish new regulations defining broken corn and broken kernels. The act provides that no dockage or foreign material once removed from grain can be recombined with any grain. Dust is prohibited from being re-introduced to the grain stream at any location.

Insect infestations present a serious problem in grain quality; they are one cause of foreign and domestic complaints. The administrator of the Federal Grain Inspection Service (FGIS) is required to make grain inspection standards that will reflect insect infestation more accurately and evaluate the effects of adopting a new optional grain grading system. The FGIS is required to adopt, to the extent practicable, the tests that are economically feasible. The FGIS administrator is required to submit (1) an annual report to Congress on the ongoing review, and (2) report to Congress on the results of the study and actions taken not later then one year after enactment.

The 1986 act will not be the last word in improving grain quality, but if its provisions are practicable and are followed consistently, it can be a major step in improving the market for U.S. grain exports.

Transforming Farm Financial Policy

The years from 1972 to the end of 1982 saw the greatest growth in farm debt and appreciation of capital assets, especially land, ever experienced. In the spring of 1983, land prices dropped sharply. Evidence of financial stress became more pronounced, as net farm income dropped sharply. In the first half of the 1980s, the total equity value of the American farm (total assets minus total debts) dropped by 25 percent. Farmers experienced declining ability to service debt out of current earnings. By 1986, 20 to 25 percent of farm proprietors were in financial difficulty.[13]

From the early 1970s through the early 1980s, interest payments on farm debt rose by nearly 260 percent, and reflected the rise in debt outstanding of more than 60 percent in real terms (1982 dollars) and the higher interest rates of the early 1980s. The sharp decline in interest rates in the mid-1980s was a boon to farmers who could refinance, but the combination of weak or non-performing farm loans and lower interest rates affected all farm lenders. Those lenders depending most heavily on farm debt payments, such as the federal land banks and the production credit associations (PCAs), were hurt most of all. Proposals to refinance farm debts and scale down or stretch out debts could not be met by these lenders without considerable help from the federal government. A rippling effect spread the consequences of these problems among all farm lenders, agribusinesses, and rural communities.

A study involving more than 800 simulations of individual farm financial situations, by 30 agricultural economists in 14 major farm states, specified a need to target the development and use of public assistance programs to the financial characteristics of the intended recipients. For instance, different policy options considered included the following: (1) sales of the farm's assets with a leaseback of these assets by the farm operator (sales were at current market values with proceeds applied against indebtedness); (2) sales of the farm's assets without leaseback; (3) writedowns of the farm's initial indebtedness; (4) reductions in interest rates on the farm's indebtedness; (5) deferrals of scheduled payments of principal and interest on intermediate-term debt for two years with no accrual of interest in the interim; and (6) infusions of outside equity capital to displace the farm's indebtedness.[14]

The possibilities for successful refinancing vary widely among farmers in differing financial condition. In general, those farmers with a debt-to-asset ratio under 40 percent and a positive return on assets can survive and generally prosper without assistance. Those with a debt-to-asset ratio of 40 to 70 percent and a positive return on assets or equity of five percent or more on an equity above $50,000 are generally classed as in good condition. Those with a debt-to-asset ratio of 70 percent or more and without a return on equity above five percent will be classed as stressed, or vulnerable. Generally, the farmers in an intermediate position in respect to debt and return on equity can benefit the most from a public assistance program.[15] (Also see discussion in Chapter 1.)

What public action should be taken to refinance farms in financial difficulty? On the one hand, the only responsibility of the federal government is to ensure the financial integrity of the institutions under its policy control--the federal land banks, the intermediate credit banks, the banks for cooperatives, and the federally-financed Farmers Home Administration. However, there are precedents for saying that the U.S. government should step in with new financing programs to refinance most of the farm operators experiencing financial difficulty.

The Farm Credit Act of 1971 greatly expanded the lending power of the federal land banks and thus enabled them to expand their loans to farmers from about $7.1 billion at the beginning of 1971, not quite one-quarter of the $29.5 billion of farm mortgages outstanding, to about $42.0 billion by 1982, more than two-fifths of the $100 billion of farm mortgages outstanding. Although all other farm mortgage lenders increased their loans, from $22.4 billion in 1971 to about $58.0 billion in 1982, the great increase in farm credit was led by the federal land banks.

Production credit associations (PCAs), financed by the intermediate credit banks, and the Farmers Home Administration were the leaders in expanding non-real estate farm debt from 1971 to 1982. Although commercial banks tripled the total of their non-real estate farm lending during these years, their growth lagged behind the growth of the federally sponsored agencies.

What is the moral responsibility of the federal government to come to the aid of financially distressed farmers? Hard choices will have to be made concerning whether to help farm proprietors in financial difficulty, under what circumstances to help them, and by how much. For instance, if there are 500,000 farms in financial difficulty should it be the goal of government to (a) let them fail, (b) save 200,000 by refinancing, or (c) save nearly all by a massive refinancing program?

Many farmers, who are in sound financial condition, believe that it is unfair or unsound for the government to rescue those who have been imprudent in borrowing and investing. Yet many of the short-term problems of the American farm are directly the result of government-induced borrowing. There is precedent for government help, such as occurred in the 1930s, but there is no easy answer to the questions of how much and under what circumstances help should be given.

It is important, however, in any type of rescue operation, to try to treat the problem, rather than just the symptoms. If a farmer is so far in debt that current interest and amortization payments cannot be met, it is important to determine whether refinancing will help the farmer to recover, or whether it will merely postpone the time of liquidation or bankruptcy. Firms that lack long-term profitability and cannot change the nature and scope of their operations sufficiently to assure profitability are poor candidates for credit subsidies or refinancing.

Apart from refinancing of individual farm firms and creating conditions for more general recovery of the farm economy, lending institutions might be encouraged to take title to farm property in lieu of debt obligations, and then lease the property back to the original debtor or another operator.

Without substantial improvement in farm incomes or decreases in interest rates charged to farmers, none of the short-term credit rescue policies holds much promise over such periods as a decade. Fundamental help for the long term is a function of market recovery, which may be helped by changes in farm commodity policy, growth in the world economy, and a fiscal-monetary policy that does not penalize the American farm.

Transforming Soil Conservation Programs

The chief problem in designing and implementing soil conservation programs is that soil erosion is a relatively minor cost in most farming situations while the costs of off-site damages are substantially more. Although research shows that soil erosion is increasing, and, hence, off-site damage costs are also increasing, continuing present erosion rates would not likely impose significantly higher costs of food and fiber on future generations. Many farmers do a good job of controlling erosion, but it is seldom in their best economic interest to farm so that the off-site damage costs will be minimized.

Soil erosion is highly concentrated in a few crop-growing regions of the United States, yet soil-conservation funds have always been widely dispersed. Farmers wish to receive their share of available funds, and wide sharing creates political support for programs. Two-thirds of the soil erosion in the United States occurs on crop land, and crop land erosion is also concentrated geographically. For instance, on a per acre basis the major problems are in the Corn Belt, Appalachia, Southern Plains, and the Mountain States. Most erosion is due to water run-off, but wind erosion is especially important in the Southern Plains and the West.

Soil erosion is also highly concentrated within regions, within individual states, sometimes within counties, and sometimes even within individual fields on a farm. The western corn belt states of Illinois, Iowa, and Missouri have 16 percent of the nation's crop land and 31 percent of its soil erosion. Within each of these states erosion tends to be concentrated among those individual watersheds that are cropped and more erosive than other areas.

Economists, soil scientists, farm managers, and other professionals recognize that to reduce costs of off-site damages of erosion to more acceptable levels, more control measures must be concentrated where erosion occurs and damages are most severe. Traditionally, people have thought of soil conservation programs as primarily for the benefit of farmers to maintain the productivity of their land which in turn would be of benefit to society. Now the main problem is to reduce and control the off-site damage of stream and reservoir sedimentation, downstream flooding, pollution of water systems, destruction of wildlife habitats, and other unwanted environmental effects.

To do this requires a new system for appraisal of the sources of the damages and for targeting the funds and other resources for erosion control. It is not appropriate to have a general goal of reducing erosion on all crop land to a uniform tonnage level such as five tons per acre, which has been a goal in the past. In some instances a higher level of soil loss can be tolerated, perhaps indefinitely. Some farm land that has a high erosion level, such as 10 or more tons of soil loss per year, should be taken out of crop production. This could be done by new state laws, or by purchase or long-time leases by the federal government. On other lands, states may have to impose new types of zoning to control erosion on lands that are highly erosive when cropped, but still profitable for other farm uses.

Transforming Federal Crop Insurance[16]

The main continuing problem in crop insurance is how best to protect farmers against yield losses due primarily to adverse weather or other conditions over which the farmer has little or no control. The basic concept is that crop insurance can be offered as a contract between the federal government and the farmer to protect the farmer's income against the risks and uncertainties of weather but not against low yields resulting from bad judgment. The former are computable as in a frequency distribution, and hence insurable. The latter will vary from farmer to farmer and from season to season and cannot be accepted as an insurable risk without considerable cost to the insuring agent, or high premiums for the insured.

Federal crop insurance has been designed to insure the actual yields of the individual farmers, rather than the conditions of crop production. This design has resulted in generally low participation by farmers, because of high premiums, and rather large net costs to the federal government in most years, due to adverse selectivity and uncomputable crop losses.

The general policy goal has been to assure a given level of crop return and to provide for greater stability in land tenure and farming operations. Over the years, however, the FCIC program has added only a very modest amount of income stability to the crops economy.

Farmers have indicated a continuing interest in crop insurance, but the majority have said that the levels of coverage have been too low, relative to the premiums charged. Over the years, most farmers have indicated by their behavior that they believe they are better off without crop insurance. Sample studies have revealed that as many as half of the farmers who buy insurance do so in lieu of greater diversification of crops and livestock.

Because of the high cost of disaster payment programs in the 1970s, however, the federal crop insurance program was expanded in the early 1980s, with the goal of making it the nation's primary means of disaster protection for farmers. The Federal Crop Insurance Act of 1980 authorized up to 30 percent of the crop insurance premium to be paid by the federal government. It also provided for higher yield guarantees and larger indemnities for each bushel or pound of crop loss, and expanded the program to cover several more crops and many more counties than had been served previously. Private insurance companies were authorized to serve as agents.

By 1983, a new liberalized Individual Yield Coverage plan (IYC) had been developed that would enable top producers to qualify for higher production guarantees without increasing their per-acre insurance cost. Farmers could choose to insure their crop for 50 percent, 65 percent, or 75 percent of average yield, and the federal government would pay 30 percent of the premium up to the 65 percent coverage level. With the federal subsidy, the FCIC's goal was a 90 percent loss pay-out, with 10 percent added to a reserve for catastrophic losses. Indemnity for losses has been determined on a unit basis, with payment being made when production was less than the guarantee, units of production being converted to dollars at pre-selected prices, and premiums varying directly with the level of insurance selected by the farmer.

Federal crop insurance could be transformed into an actuarily sound system by taking these provisions a step further to base indemnities on the yield experience in a good pre-selected sample of farmers in a community, or an area in which a certain type of farming prevails. In case of low yields in a community or crop area, all farmers who were insured for a yield level higher than the average yield experienced in that year would be paid an indemnity. A farmer who had unusually low yields due to poor farming, neglect or theft, would not receive an additional indemnity and a farmer who had unusually high yields would not be denied an indemnity because of superior performance. Additional factors such as average rainfall, frost-free days, or average temperature might be taken into account in appraising whether or not the average yield of the community or crop area accurately reflected the yield conditions faced by farmers in the community or area. Separate hail insurance provisions could be added to the basic contract. By such contracts, crop insurance could be transformed into a more actuarily sound system of more reliable use to farmers.

A General Conclusion

There are a number of options for transforming farm policy into a system that will benefit both farmers and the general public more. The key policy decision rests in the area of commodity programs. Government intervention has not worked the way it was intended by those who have favored price support and production control programs. The option exists for phasing out most of these programs with various alternatives for compensating farmers during a transition period. The domestic and world market situation places the American farm in a favorable position to take advantage of the new age of high technology. Additionally, new options exist for improving the quality of U.S. farm exports, stabilizing farm financial policy, transforming soil conservation programs, and improving the actuarial soundness of federal crop insurance.

A broad survey of the literature shows that many agricultural economists and other professionals agree on the goals of farm policy and the means to achieve these goals. Whether the goals are achieved depends on policy decisions and actions that may be taken.

Endnotes

1. Dennis Avery. "U.S. Farm Dilemma: The Global Bad News Is Wrong," *Science* 230 (October 1985). p.412.

2. Alex McCalla. "Assessment from Outside the Beltway," *U.S. Agricultural Policy: The 1985 Farm Legislation*. Washington, DC: American Enterprise Institute, 1985. p.193.

3. United States. Economic Research Service. *Dairy Situation and Outlook Report*. DS-405 (June 1986), DS-406 (July 1986), and DS-408 (December 1986). Washington, DC: U.S. Government Printing Office, 1986.

4. For current projections see the *Population Bulletin* (published quarterly), or United States Bureau of the Census, *Demographic Estimates for Countries with a Population of 10 Million or More* (published annually).

5. John W. Mellor. *The New Global Context for Agricultural Research: Implications for Policy. Food Policy Statement*, no. 6. Washington, DC: International Food Policy Research Institute, 1986.

6. D. Smallwood and J. Blaylock. *Impact of Household Size and Income on Food Spending Patterns*. Technical Bulletin, no. 1650. Washington, DC: Economic Research Service, U.S. Department of Agriculture, 1981.

7. J. Kinsey, ed. *Consumer Demand and Welfare: Implications for Food and Agricultural Policy*. North Central Region Publication, no. 311. St. Paul: Agricultural Experiment Station, University of Minnesota, 1986.

8. D. Smallwood, J. Blaylock, B. Sexauer, and J.S. Mann. *Food Expenditure Patterns of Single Person Households*. Agricultural Economics Report, no. 428. Washington, DC: Economic Statistics and Cooperative Service, U.S. Department of Agriculture, 1979.

9. The Joint Nutrition Monitoring Evaluation Committee of the United States Department of Agriculture and the Department of Health and Human Services. *Nutrition Monitoring in the U.S.: Progress Report of the Joint Nutrition Monitoring Evaluation Committee*. Washington, DC: U.S. Government Printing Office, 1986.

10. United States. Department of Agriculture. *Food Consumption and Expenditure Statistical Bulletins*, nos. 565 and 736, and *Background for Farm Legislation Bulletins*, nos. 466 and 476. Washington, DC: U.S. Government Printing Office.

11. E. Ned Sloan. *Preliminary Review of Grain Quality Improvement Act, 1986*. FG15. Washington, DC: U.S. Department of Agriculture, 1986.

12. Lowell D. Hill, Marvin Paulsen, and Margaret Early. *Corn Quality: Changes During Export*. Urbana: Agricultural Experiment Station, College of Agriculture, University of Illinois, 1979. Lowell D. Hill. "Principles for Use in Evaluating Present and Future Grain Grades." Department of Agricultural Economics, University of Illinois (September 1985), no. 85 E-329; "Regulation and Economic Incentives for Improving Grain Quality," no. 86 E-357; and Statement of Lowell D. Hill before the subcommittees on: Wheat, Soybeans, and Feed Grains and Department Operations, Research and Foreign Agriculture, U.S. House of Representatives. Committee on Agriculture

Related to Grain Standards. Urbana: University of Illinois at Urbana-Champaign, 1986.

13. *The Farm Credit Crisis: Policy Options and Consequences*. AE-4612. Urbana: College of Agriculture Cooperative Extension Service, University of Illinois at Urbana-Champaign, 1986. Eight articles by 17 economists from the federal government and 9 universities discuss the farm credit crisis and alternative policy options.

14. From an unpublished paper by Peter J. Barry, Paul N. Ellinger, and Vernon R. Eidman. "Farm Level Adjustments to Financial Stress," 1986.

15. Peter J. Barry. *Financial Stress in Agriculture: Policy and Financial Consequences for Farmers*. A research activity associated with Southern Regional Research Project S-180, An Economic Analysis of Risk Management Strategies for Agricultural Production Firms. Urbana: Department of Agricultural Economics, Agricultural Experiment Station, College of Agriculture, University of Illinois at Urbana-Champaign, 1986.

16. This section draws on a series of studies by Harold G. Halcrow. "The Theory of Crop Insurance." Ph.D. dissertation, The University of Chicago, 1948; "Actuarial Structures for Crop Insurance," *Journal of Farm Economics* 31:3 (August 1949); *Agricultural Policy of the United States*. Englewood Cliffs, NJ: Prentice-Hall, 1953. pp. 407-420; *Food Policy for America*. New York: McGraw-Hill, 1977. pp. 425-429; and "A New Proposal for Federal Crop Insurance," *Illinois Agricultural Economics* 18:2 (July 1978). pp.20-29. The last study was done at the request of the United States Senate Committee on Agriculture, Nutrition and Forestry.

Chapter 7: Annotated Bibliography

Commodity Program Alternatives

*** 7.1 ***

Alaouze, Chris M., A.S. Watson, and N.H. Sturgess. "Oligopoly Pricing in the World Wheat Market." *American Journal of Agricultural Economics* 60:2 (May 1978):173-185.

The authors examine the hypothesis that "When the residual demand curve for wheat facing the United States and Canada shifts to the left, or when the exportable surplus of Australia is large, market-shares of these duopolists are reduced. Such circumstances lead to the formation of a market-share triopoly with Australia." A model of triopoly pricing in the world wheat market is presented and the authors conclude that "If major exporters continue to be concerned with relative market-shares, the triopoly will reform, stocks will accumulate, and lower prices will prevail; however, prices will be more variable, and possibly higher, than before 1972/73."

*** 7.2 ***

Babcock, Bruce, and Andrew Schmitz. "Look for Hidden Costs: Why Direct Subsidy Can Cost Us Less (and Benefit Us More) Than a 'No Cost' Trade Barrier." *Choices: The Magazine of Food, Farm, and Resource Issues* (Fourth Quarter 1986):18-21.

Babcock and Schmitz argue that "it is misleading to judge the relative merits of government programs simply on the basis of taxpayer expense. The effects of policy on society's welfare are broad and complicated. One must identify and quantify both consumer and producer gains and losses from government intervention before choosing the best policy." For instance, deficiency payment schemes as we have for corn, wheat, rice, and cotton "tend to encourage greater production and so result in food prices lower than they would be without the program. In such cases the gains to consumers from the marketplace offset the burdens of tax expenditures." The sugar program, in contrast, costs the government very little; but as Babcock and Schmitz estimate "When we buy sugar at the supermarket, or purchase any of the hundreds and thousands of products containing sugar, we pay an extra 16 cents for each pound of sugar."

*** 7.3 ***

Bookins, Carol. "The Embargo Study: Domestic Policy Response Caused Long Term Damage." *Choices: The Magazine of Food, Farm, and Resource Issues* (First Quarter 1987):21.

The conclusions of the embargo study released in November 1986 have caused much controversy. Here Bookins contends that "the study failed to emphasize the effect of domestic agricultural policy changes implemented as a result of embargoes...the political crisis management of the 1980 embargo did lead to domestic policies which heightened distortions in the world agricultural economy. A conclusion on which all can perhaps agree is that the political economy of American agriculture is an area that warrants much greater study."

*** 7.4 ***

Brada, Josef C. "The Soviet-American Grain Agreement and the National Interest." *American Journal of Agricultural Economics* 65:4 (November 1983): 651-656.

In the 1970s the Soviet Union became a major importer of grain, which gave rise to fears that it could extract an undue share of the gains from East-West trade through the monopsonistic power of its state-trading organs and by keeping its buying intentions secret. Brada constructed offer curves depicting Soviet trade using two alternative characterizations of Soviet trade decision making. He showed that "the grain agreement between the two countries negates cost of the bargaining advantage attributed to the Soviet trade monopoly and may open the Soviet Union to exploitation by the United States."

*** 7.5 ***

Carman, Hoy F. "A Trend Projection of High Fructose Corn Syrup Substitution for Sugar." *American Journal of Agricultural Economics* 64:4 (November 1982):625-633.

High fructose corn syrup is a comparatively low-priced sugar substitute experiencing rapid demand growth. Carman constructed a simple logistical trend model that suggested that both per capita and total U.S. sugar consumption will decrease for several years as high fructose corn syrup is adopted. He concluded, however, that "The impact on domestic sugar producers under current policy is minimal with costs borne primarily by sugar-exporting countries. Food manufacturers can reduce production costs, and some of this may be passed to consumers in lower product prices. Per capita consumer savings, however, will be small. The impact on corn prices also will be small."

*** 7.6 ***

Chambers, Robert G., and Richard E. Just. "Effects of Exchange Rate Changes on U.S. Agriculture: A Dynamic Analysis." *American Journal of Agricultural Economics* 63:1 (February 1981):32-46.

Chambers and Just used an econometric model of the wheat, corn, and soybean markets to examine the dynamic effects of exchange rate fluctuation on U.S. commodity markets. Exports and agricultural prices were found to be sensitive to movements in the exchange rate, while domestic factors, such as disappearance and inventories, were less sensitive but still responsive. Dramatic short-run adjustments in prices and exports were followed by less dramatic but significant longer-run adjustments. They concluded that "the hypothesis of elastic response to the exchange rate seems particularly relevant for the short run."

*** 7.7 ***

Clarkson, Kenneth W. *Food Stamps and Nutrition.* Washington, DC: American Enterprise Institute for Public Policy Research, 1975. ISBN 0-8447-3155-2.

This study is highly critical of the food stamp program in respect to both waste in the program and certain failures to improve nutrition through use of stamps. The study offers important background information for reviewing current food stamp problems and improving the program.

*** 7.8 ***

Cochrane, Willard W. "The Need to Rethink Agricultural Policy in General and to Perform Some Radical Surgery on Commodity Programs in Particular." *American Journal of Agricultural Economics* 67:5 (December 1985):1002-1009.

Cochrane, who for many years strongly advocated strict production controls for the American farm, argues that price supports and production controls on major farm commodities should be discontinued, primarily because they are "helping one group of farmers do in another group." He would, however, "keep government in agriculture where it operates to make the competitive game of farming more fair to all concerned, hence, a more acceptable game for all concerned."

*** 7.9 ***

United States. Congress. Congressional Budget Office. *Diversity in Crop Farming: Its Meaning for Income-Support Policy.* Special Study. Washington, DC: U.S. Government Printing Office, 1985.

A principal purpose of the federal government's farm programs is to protect crop farmers from losses of income when prices fall. These production-based farm programs are not designed to provide income support on the basis of need. Among those who do benefit, the diversity of crop farming causes the programs to have uneven effects on farm incomes. The study concludes that "A policy of targeting income support to low-income farmers could be combined with the present policy of stabilizing prices and incomes for all farmers. Many farm families do not have chronically low incomes, but they operate under conditions of risk and uncertainty so that their incomes are highly variable. Public policy has long acknowledged this fact. Targeted income support to low-income farmers need not conflict with those policies aimed at stabilizing the incomes of other farmers."

*** 7.10 ***

Frankel, Jeffrey A. "Expectations and Commodity Price Dynamics: The Overshooting Model." *American Journal of Agricultural Economics* 68:2 (May 1986):344-347.

Frankel notes that monetary policy has important effects on farm commodity prices because, though they are flexible, other goods prices are sticky. His paper formalizes the argument by applying the Dornbusch overshooting model, in which a decline in the nominal money supply is a decline in the real money supply in the short run. Such a decline raises the real interest rate, which depresses real commodity prices. Frankel shows that commodity prices overshoot their new equilibrium in order to generate an expectation of future appreciation sufficient to offset the higher interest rate. These real effects (which vanish in the long run) also result from a decline in the money growth rate.

*** 7.11 ***
Gardner, Bruce L. *The Governing of Agriculture.* Lawrence: The Regents Press of Kansas, 1981. ISBN 0700602143.

Gardner's theme is never in doubt. "The search for solution to agriculture's problems through governmental intervention has been and will continue to be a costly delusion." The solution? Phase out current farm commodity programs and ease into market approaches. In a careful, scholarly, and relatively short analysis, Gardner concludes that the nation's policies in dealing with the farm commodity markets do not promote the public interest as they should and that the principal route to improvement is through less intervention in these markets.

*** 7.12 ***
Hathaway, Dale E. "Trade Negotiations: They Won't Solve Agriculture's Problems." *Choices: The Magazine of Food, Farm, and Resource Issues* (Fourth Quarter 1986):14-17.

Hathaway argues that "It is a mistake for U.S. agricultural groups to believe that even a successful MTN (Multilateral Trade Negotiations) in agriculture will solve most of their problems." His reasons are that "The MTN is not the forum for dealing with issues of expanding world economic growth and world demand for agricultural products,...the matter of stabilizing exchange rates and coordinating national economic policies, [or].... What can be done internationally to deal with the current excess capacity in world agriculture." Currently, "Market forces are not being allowed to remove this excess capacity." He concludes that an MTN in agriculture is not a bad idea, but broader solutions are required for the times ahead.

*** 7.13 ***
Heady, Earl O., and Stanley A. Schraufnagel. "An Alternative to Land for Supply Control." *North Central Journal of Agricultural Economics* 3:1 (January 1981):21-27.

"This article examines the pattern of production and resource use under two agricultural supply control programs. One is a land set-aside program and the other is a fertilizer reduction program. The study was made in light of recent changes in energy supplies and prices to determine whether supply control programs resting on inputs other than land might be more consistent with developing conditions of resource endowments. The tool of analysis is a national and interregional programming model. Results indicate that, among other things, less energy would be used and more soil might be lost to erosion under the fertilizer program than under the land program."

*** 7.14 ***
Helms, L. Jay. "Errors in the Numerical Assessment of the Benefits of Price Stabilization." *American Journal of Agricultural Economics* 67:1 (February 1985):93-100.

Helms presents a simple computational procedure for determining a consumer's willingness to pay to have a price stabilization policy implemented. Numerical simulations are then provided, which demonstrate that the commonly used expected surplus measures (whether Hicksian or Marshallian) can in fact seriously misstate the true benefits of price stabilization. These simulations also illustrate the extent to which the results may be dependent upon the assumptions made--implicitly or explicitly--regarding the consumer's attitudes toward risk, thereby underscoring the need to conduct sensitivity analyses to determine the robustness of benefit-cost assessments with respect to these assumptions. Helms concludes that "In assessing the welfare implications of price stabilization policies, true measures of the consumer's willingness to pay to have a program implemented must reflect the consumer's risk preferences as well as the dispersion of the price distribution itself."

*** 7.15 ***
Hillman, Jimmye S., ed. *United States Agricultural Policy for 1985 and Beyond.* Tucson: University of Arizona and Resources for the Future, 1984.

Twelve agricultural economists offered this series of lectures on U.S. agricultural policy for 1985 and beyond. The collection comprises seminar papers, which were presented biweekly, before an audience of faculty and graduate students in the Agricultural Economics Department and selected others at the University of Arizona during the spring of 1984.

*** 7.16 ***
Karp, Larry S., and Alex F. McCalla. "Dynamic Games and International Trade: An Application to

the World Corn Market." *American Journal of Agricultural Economics* 65:4 (November 1983):641-650.

Karp and McCalla note that dynamic games are a conceptually useful way of analyzing imperfect markets where both buyers and sellers have potential market power and that previous analysis of imperfect markets was static and limited to either exporters or importers. A dynamic game allows the inclusion of both importers and exporters in a multiperiod framework allowing the derivation of reaction functions; and they applied a Nash noncooperative difference game to the international corn market to explore the plausibility of numerical results. The authors concluded that "The results are reasonable and show that the game approach based on a more comprehensive econometric model has a promising future in policy analysis."

*** 7.17 ***

Langley, James A., Lyle P. Schertz, and Barbara C. Stucker. "Alternative Tools and Concepts." In *Agricultural-Food Policy Review: Commodity Program Perspectives.* Agricultural Economic Report, no. 530. Washington, DC: Agricultural Economic Research Service, U.S. Department of Agriculture, 1985.

As an alternative in commodity programs, the authors note that a moving average of past prices would allow loan rates to adjust to changes in market trends, yet provide a safety net for farmers. The development of options markets and the legalization of trade options would be important to the commercial viability of revenue insurance for individual crop producers. Revenue insurance would not necessarily involve income transfers to producers, but income transfers could be linked to transactions dealing with insurance of individual producers' revenue. Another alternative could be joint producer and government activities providing income "assurance" to a group of producers as contrasted to insurance for individual producers.

*** 7.18 ***

Lin, William, James Johnson, and Linda Cavin. *Farm Commodity Programs: Who Participates and Who Benefits.* Agricultural Economic Report, no. 474. Washington, DC: Economic Research Service, U.S. Department of Agriculture, 1981.

The authors show in good detail how program benefits are distributed among farms and by commodities. Direct benefits are highly skewed according to volume of farm production, and benefits are for the most part concentrated among a few commodities. Although the authors do not make recommendations, they do provide important information that may be useful in transforming commodity programs.

*** 7.19 ***

Michie, Aruna Nayyar, and Craig Jagger. *Why Farmers Protest: Kansas Farmers, the Farm Problem, and the American Agriculture Movement.* Manhattan: Agricultural Experiment Station, Kansas State University, 1980.

In this study of farm discontent in the midwest, the authors cite the major reasons for which farmers protest. Among the chief of these are a feeling of exploitation, unfair prices, and a system that farmers feel is weighted against them, all of which lead them to support protest organizations as a means of gaining power.

*** 7.20 ***

Pasour, E.C., Jr. "Cost of Production: A Defensible Basis for Agricultural Price Supports?" *American Journal of Agricultural Economics* 62:2 (May 1980): 244-248.

Pasour discusses problems of defining and measuring costs used as a basis for price supports and notes that "The cost of producing a commodity is meaningful only in a static equilibrium model and, even then, cost cannot be defined independently of demand when resources are specialized. Under real world conditions, opportunity costs are subjective and vary widely between producers." His analysis does not deny that government can vary output levels by price setting. What he challenged is the contention that such an approach involves the basing of price supports on cost. He concluded that "In view of the problems raised above, attempts to define and estimate 'reasonable cost estimates' as a basis for agricultural price supports appear to be fore-doomed."

*** 7.21 ***

Richardson, James W., and Daryll E. Ray. "Commodity Programs and Control Theory." *American Journal of Agricultural Economics* 64:1 (February 1982):28-38.

Richardson and Ray cast the decision-making process for commodity programs in the United States in the terms of adaptive control theory, following the control framework outlined by Rausser. They argued that the actual commodity program decision-making process can be viewed as a sequential multiperiod control problem and that multiple period, open-loop feedback control of a disaggregate policy simulation model can be used to assist agricultural policy makers. They concluded that their "empirical results present the first-round application of a suggested sequential use of control techniques and demonstrate the feasibility of applying control procedures to a multicommodity, agricultural policy simulator."

*** 7.22 ***

Salathe, Larry, J. Michael Price, and David E. Banker. "An Analysis of the Farmer-Owned Reserve Program, 1977-83." *American Journal of Agricultural Economics* 66:1 (February 1984):1-11.

The authors use an econometric model of the U.S. agricultural sector to examine the effects of the Farmer-Owned Reserve Program (FORP) on crop and livestock production and prices, farm income, crop carryover levels, and government outlays. They concluded that "The program has had a positive impact on commodity prices and farm income but has not significantly reduced the annual variation in commodity prices. It also increased government outlays for agricultural commodity programs, but all of the increase is potentially recoverable. The continued use of the FORP to enhance commodity prices likely will lead to excessive government outlays in the long run."

*** 7.23 ***

Shaffer, James, Vernon Sorenson, and Lawrence Libby, eds. *The Farm and Food System in Transition: Emerging Policy Issues*. A series of 50 leaflets sponsored by the Extension Committee on Policy (ECOP), USDA-Extension, Michigan State Cooperative Extension Service, and the State Cooperative Extension Services. East Lansing: Department of Agricultural Economics, Michigan State University.

The authors of these leaflets in a series provide a comprehensive discussion of the U.S. farm and food system and the related public policy issues expected to be on the agenda for the 1980s. The leaflets are planned to be used individually or as sets by readers with specific interests, and as a total collection for those seeking a general understanding of the modern U.S. farm and food system.

*** 7.24 ***

Whipple, Glen D. "An Analysis of Reconstituted Fluid Milk Pricing Policy." *American Journal of Agricultural Economics* 65:2 (May 1983):207-213.

Whipple used a reactive programming model of the U.S. milk market to simulate the effects of altered reconstituted fluid milk pricing policy. The solutions indicated that reconstituted fluid milk, as a lower cost alternative to fresh fluid milk, would make up a substantial portion of the fluid milk consumption in some markets. His analysis led to the conclusion "that alteration of the reconstituted fluid milk pricing provisions of federal and state milk market orders would have a substantial impact on market equilibrium."

World and Domestic Market Perspectives

*** 7.25 ***

Bigman, David. *Coping with Hunger: Toward a System of Food Security and Price Stabilization*. Cambridge, MA: Ballinger Publishing Co., 1982. ISBN 0884103714.

Despite its title, *Coping with Hunger* is not a survival manual for the mal-nourished or another doom-and-gloom diatribe on the global food picture, but an attempt to combine standard, simple, neoclassical tools of economics with numerical simulation to appraise different food policies and the cost effectiveness of various instruments. One should not be deterred by the inevitable air of unreality of simulation exercises, the modest amount of algebra, or the continuing saga of what is food security.

*** 7.26 ***

Bigman, David, and Shlomo Reutlinger. "National and International Policies Toward Food Security and Price Stabilization." *The American Economic Review* 69:2 (May 1979):159-163.

At a time when there was continuing concern about the adequacy of domestic and world food supplies, the conclusion of Bigman and Reutlinger was that "At current food production levels, supplies are adequate to feed the world's population even in lean years." Their basic premise was "that current food crises are characterized not by overall scarcity but by gross mal-distribution of food. With the existing distribution of income and wealth among individuals and nations, the free-market economy is unable to prevent frivolous uses of food at the same time that other people are starving. While trade and buffer stocks can play a major role in stabilizing food grain supplies, effective insurance against hunger will also require special financial measures." Their analysis showed that national food security measures by poor countries are very costly, involve substantial transfers of income, and require massive intervention by the government in the free market with possible undesirable consequences in the long run. Alternatively, or at least in addition, they proposed an international financial undertaking that would enable developing countries to acquire food in times of need.

*** 7.27 ***

Boserup, Ester. *Population and Technological Change: A Study of Long-Term Trends*. Chicago: University of Chicago Press, 1981. ISBN 0226066738.

Boserup analyzes historical and contemporary evidence on the long-term interrelations between population change and technological change in pre-agricultural communities and societies at the beginning of industrialization. She defines technological change broadly to include agricultural methods, health and sanitation techniques, administration methods, transportation techniques, and literacy. Her earlier work on this topic (1965) engendered considerable controversy among demographers because of her generally positive assessment of the effects of population growth on agricultural progress. Those who found comfort in her earlier book should take note of the many examples in this book in which rural population growth is accompanied by diminishing returns, stagnation in rural wages, and increased hours of work by rural residents.

*** 7.28 ***

Bredahl, Maury, Andrew Schmitz, and Jimmye S. Hillman. "Rent Seeking in International Trade: The Great Tomato War." *American Journal of Agricultural Economics* 69:1 (February 1987):1-10.

The authors presented a model of international rent-seeking activities by producers in both exporting and importing nations and applied it to the winter vegetable trade between the United States and Mexico. Attempts to form export/import coalitions were analyzed and reasons for certain failures were given. The authors concluded that "Due to past failures to impede trade, essentially free trade in winter vegetables between the two countries exists."

*** 7.29 ***

Brown, Lester R. *Building a Sustainable Society, a World Watch Institute Book.* New York: W.W. Norton and Company, 1981. ISBN 0393014827.

Brown has written one of the more important books in a continuing series of studies by the World Watch Institute on world resources, and future demands for resource use. The author advocates caution in the use of nonrenewable resources and outlines policies required to sustain society over the long term.

*** 7.30 ***

Brown, Peter G., and Henry Shue, eds. *Food Policy--the Responsibility of the United States in the Life and Death Choices.* New York: The Free Press, A Division of Macmillan Publishing Co., 1977. ISBN 0029049806.

The authors present a rather grim picture of the world food situation over the rest of this century and the early part of the next, writing at a time when concerns over food shortages were tending to dominate professional discussion. The United States is assigned a leading role for preventing hunger and starvation. Although increased world food production in the 1980s changed the general outlook, the issues of world responsibility remains.

*** 7.31 ***

Bryson, Reid A., and Thomas J. Murray. *Climates of Hunger: Mankind and the World's Changing Weather.* Madison: The University of Wisconsin Press, 1977. ISBN 0708110533.

Bryson and Murray have written a scholarly study of weather changes over several centuries and the changes that occur in plant and animal populations as a result of the changing weather. Although their book is not directly related to short-term policy decisions, it gives important background information for consideration of future long-term policy.

*** 7.32 ***

Campbell, Keith O. *Food for the Future, How Agriculture Can Meet the Challenge.* Lincoln: University of Nebraska Press, 1979. ISBN 0803209657.

Campbell presented a generally optimistic view of future world food supplies at a time when many professionals were visualizing a more limited or grim future. Fortunately, Campbell has been proven correct, as far as total world food production is concerned. But serious problems of food distribution and related issues remain to be solved.

*** 7.33 ***

Chambers, Robert G., and Michael W. Woolverton. "Wheat Cartelization and Domestic Markets." *American Journal of Agricultural Economics* 62:4 (November 1980):629-638.

The authors analyzed the effects of cartelization in the international wheat market, giving particular attention to the relationship between the level of cartel profits and the price farmers receive for their wheat. Sufficient conditions were established for a rise in cartel profits to be associated with a rise in the farm gate price of wheat. They also investigated the potential stability of a cartel formed by the major trading companies and introduced and discussed a means for preventing cheating.

*** 7.34 ***

Crosson, Pierre, and Kenneth D. Frederick. *The World Food Situation.* Washington, DC: Resources for the Future, 1977. ISBN 0-8018-2062-6.

In a well-documented study, Crosson and Frederick emphasize the importance of natural resource conservation in protecting the productivity of farming in the United States and around the world.

*** 7.35 ***

Grigsby, S. Elaine, and Cathy L. Jabara. "Agricultural Export Programs and U.S. Agricultural Policy." In *Agricultural-Food Policy Review: Commodity Program Perspectives.* Agricultural Economic Report, no. 530. Washington, DC: Economic Research Service, U.S. Department of Agriculture, 1985.

Various programs to increase farm exports have been used in combination with domestic commodity programs as part of an overall farm policy rather than as an explicit trade policy. Declines in U.S. exports have generated increased interest in a redirection of U.S. policy to improve export performance. The authors conclude that "Even under a more market-oriented policy, export policy instruments which expand long-term demand play a role in formulation of a trade strategy."

*** 7.36 ***
Johnson, D. Gale, and G. Edward Schuh, eds. *The Role of Markets in the World Food Economy*. Boulder, CO: Westview Press, 1983. ISBN 0865316201.

Eight edited papers and attendant discussion provide "a basis for a continuing dialogue among those who have concerns about the future well-being of half the world's population that lives in the low-income developing countries of the world" (p.ix). With occasional dissent, the main theme of the papers and discussion is well-captured in a comment of one of the discussants: "What emerges from this discussion is the vital role that agricultural prices play in achieving optimum output, growth, trade, and food security. When governments intervene in the legitimate functioning of markets--even when the actions are well intentioned--the ultimate outcome is to reduce national and global welfare" (p. 141).

*** 7.37 ***
Knapp, Keith C. "Optimal Grain Carryovers in Open Economies: A Graphical Analysis." *American Journal of Agricultural Economics* 64:2 (May 1982):197-204.

Knapp uses dynamic programming to graphically analyze optimal storage, trade, and borrowing policies for grain. World prices and domestic harvests are assumed to be stochastic. Optimal carryovers are generally increasing in supply prices and decreasing in world prices, while net imports are decreasing in both supply and world prices. Optimal levels of carryovers and net imports are influenced by the opportunity to borrow and save foreign exchange. Knapp shows that grain reserves and foreign exchange borrowing/saving are alternative inventory systems that both complement and substitute for one another.

*** 7.38 ***
Newbery, David M.G., and Joseph E. Stiglitz. *The Theory of Commodity Price Stabilization: A Study in the Economics of Risk*. Oxford: Clarendon Press, 1981. ISBN 0198284179.

The authors have undertaken an ambitious task to communicate what is generally difficult subject matter to three distinct audiences: economic theorists, agricultural economists, and economic policy makers. While the book is primarily one in economic theory, and therefore of greatest interest to economic theorists, including agricultural economists, it emphasizes, where appropriate, the relevant policy conclusions to be drawn. It is mandatory reading for researchers working in the areas of risk and price stabilization. General readers will benefit from the authors' skillful treatment of policy issues.

*** 7.39 ***
Otsuka, Keijiro, and Yujiro Hayami. "Goals and Consequences of Rice Policy in Japan, 1965-80." *American Journal of Agricultural Economics* 67:3 (August 1985):529-538.

The authors use a partial equilibrium framework to analyze the change in welfare of producers and consumers, government cost, and the deadweight loss arising from the various forms of government interventions into the rice market in order to protect domestic producers in Japan. Results of the quantitative analysis indicate that the motivation of the government was to minimize budget costs in achieving the target level of producer price support, while consumer welfare had an insignificant weight in the government's objective function. The authors also found that acreage control is a "second best" policy to reduce social inefficiency produced from other forms of market distortions.

*** 7.40 ***
Paarlberg, Philip L., and Philip C. Abbott. "Collusive Behavior by Exporting Countries in World Wheat Trade." *North Central Journal of Agricultural Economics* 9:1 (January 1987):13-28.

Paarlberg and Abbott note that collusion by major grain exporters has been proposed as a means to counteract the market power of importers, thereby improving welfare. They analyze collusion by the United States, Canada, and Australia in wheat trade when both importers and exporters exercise market power by endogenously determining domestic and trade policies based on interest group lobbying. They conclude that "while each exporter realizes net gains in the short-run, the gains are uneven. Australia and Canada benefit much more than the United States."

*** 7.41 ***
Paddock, William, and Elizabeth Paddock. *We Don't Know How*. Ames: Iowa State University Press, 1973. ISBN 0-8138-1750-1.

In this noteworthy study, the authors document a number of failures and reasons for failure in programs for economic development, most of which were sponsored by the U.S. Agency for Economic Development (AID). The Paddocks maintained that these failures were due to lack of understanding of what needed to be done and they suggested a number of changes, some of which have been made.

*** 7.42 ***

Paddock, William, and Paul Paddock. *Famine-1975: America's Decision: Who Will Survive?* Boston: Little, Brown and Company, 1967. LC 67-14456.

The Paddocks assumed there would be a growing worldwide shortage of food, which occurred in a more modest way in the food crisis of the early 1970s, but which was turned around by the increasing productivity of the 1980s. The book is cited because it illustrates a type of hysteria that has occurred periodically in Western literature since Thomas Malthus wrote his famous essay on population in 1798. The Paddock's book also helps to visualize some of the problems the United States may face in the future if U.S. policymakers accept more responsibility for world food supplies than can be accommodated.

*** 7.43 ***

Schmitz, Andrew, Alex F. McCalla, Donald O. Mitchell, and Colin A. Carter. *Grain Export Cartels.* Cambridge, MA: Ballinger Publishing Co., 1981. LC 81-3448.

The authors discuss the economics of alternative wheat and coarse grain cartels that could be formed by the world's major grain exporters. They argue that a cartel could result in significant U.S. economic benefits including price stability, reduced import barriers, higher export earnings, and lower domestic grain prices. They assume that producer cartels maximize only producer welfare by setting price above the free-trade price for all users and they use surplus concepts without definition or description of their merits. Expected import demand elasticities offered by the authors are generally low, suggesting limited importer response to a cartel. But, effects of importer retaliatory alternatives are not known or described and little is known about changes in foreign cropland bases, yields, and elasticities under sustained high prices. Such changes could significantly alter cartel gains.

*** 7.44 ***

Schmitz, Andrew, Dale Sigurdson, and Otto Doering. "Domestic Farm Policy and the Gains from Trade." *American Journal of Agricultural Economics* 86:4 (November 1986):820-827.

The authors assess the extent to which the gains from agricultural trade are influenced by both U.S. domestic policy and tariff and nontariff trade barriers. The paper shows theoretically that the volume of trade can be substantial but the gains nonexistent. The authors conclude that "Empirical results, along with rent-seeking arguments, support the no-gains-from-trade hypothesis."

*** 7.45 ***

Svedberg, Peter. "World Food Self Sufficiency and Meat Consumption." *American Journal of Agricultural Economics* 60:4 (November 1978):661-666.

One of the resolutions adopted at the World Food Conference in Rome urged the rich countries to adhere to simpler and less "calorie-wasting" food consumption habits. The argument was that in this way large quantities of grain would be released for the benefit of starving people in the Third World. Svedberg, however, shows that the link between food consumption in the rich countries and the food problem of the poor countries was considered in an erroneous time perspective and that simpler food consumption habits in rich countries would do very little, if anything, to relieve the long-run food shortage in the poor countries. He concluded that "a temporary (enforced) cutback of food consumption in the rich countries may under certain circumstances be warranted to solve the short-run problem, i.e., to avert famines in the Third World in years of global crop failures." But ..."altered food habits in the rich countries seem to be an inefficient, if not impossible, means of solving the long-run food problem in the Third World. This is because (a) the decrease in food prices on the world market tends to be small because production is also going to fall, (b) the effect is not aimed directly at those starving, i.e., the lower prices will not benefit only the starving but also the rich in all countries, and (c) there is no guarantee that the governments in the starvation-stricken countries would use the additional incomes to improve the lot of the people suffering from extreme hunger."

*** 7.46 ***

Swinbank, Alan. "European Community Agriculture and the World Market." *American Journal of Agricultural Economics* 62:3 (August 1980):426-433.

Swinbank argues that the European Community's common agricultural policy is more complex than some studies would indicate. Not only do the member states succeed in maintaining nationally preferred price support levels through the use of green currencies and monetary compensatory amounts, but the protective mechanisms applied have a different impact on some commodities and some supplying countries. The introduction, in 1979, of the European Monetary System had repercussions for the agricultural sector, including the use of a new unit of account.

Improving the Quality of Grain Exports

* 7.47 *

Hill, Lowell D. *Principles for Use in Evaluating Present and Future Grain Grades*. No. 85 E-329. Urbana: Department of Agricultural Economics, University of Illinois, 1985.

In a general study, Hill outlined the basic principles for use in evaluating the impact of the U.S. government's grading system on grain exports. It also includes suggestions for changing the system to improve the quality of U.S. grain exports.

* 7.48 *

Hill, Lowell D. "Statement of Lowell D. Hill before the subcommittees on Wheat, Soybeans, and Feed Grains and Department Operations, Research and Foreign Agriculture, U.S. House of Representatives Committee on Agriculture related to grain standards." Urbana: University of Illinois, 1986.

In his statement, Hill briefly outlined the weaknesses of the existing grain grading standards and showed how this system of grading often led to inferior quality in grain exports. By allowing grain to be shipped at near the maximum levels of moisture and percentage of foreign material permitted by U.S. grain standards, grain often deteriorated after leaving U.S. ports and arrived in poor condition at ports of entry in the importing countries. Hill made a number of suggestions for Congressional action to change the grading system and indicated some of the research still needed to support such action.

* 7.49 *

Hill, Lowell D., Marvin Paulsen, and Margaret Early. *Corn Quality: Changes during Export*. Special Publication, no. 58. Urbana: University of Illinois, 1979.

In a comprehensive study, Hill and others analyze the changes that occur in the quality of corn as it goes through export channels. Corn that is dried at a high temperature tends to crack in a series of unloadings and reloadings, which leads to insect infestations or other attributes of poor keeping quality. The resulting dust and cracked kernels, added to the initially permitted foreign material and moisture, increases the total percentage of unwanted material, and also increases the likelihood of an export shipment going out of condition. Although a high percentage of shipments have arrived in importing countries in acceptable condition, the authors conclude that improvements must be made in the system to achieve more satisfactory results, and they make a number of suggestions for doing this.

* 7.50 *

Sloan, E. Ned. *Preliminary Review of Grain Quality Improvement Act, 1986*. FG 15. Washington, DC: U.S. Department of Agriculture, 1986.

Sloan, in a relative brief review of the Grain Quality Improvement Act of 1986, emphasizes the features of the act that may help to assure more uniformly high quality in U.S. grain exports. The review gives the schedule for a continuing evaluation of performance under the act, some of the needs for further research are briefly noted, and the possibility of further legislation is suggested.

* 7.51 *

United States. Congress. General Accounting Office. *U.S. Grain Exports: Concerns about Quality*. Report to the Honorable Byron L. Dorgan, House of Representatives. GAO/REED-86-134. Washington, DC: U.S. Government Printing Office, 1986.

A purpose of this study was to examine the U.S. Department of Agriculture's system for receiving and reporting on grain quality complaints, particularly the number and types of foreign complaints being received and the trend of those complaints over the past several years. It found that the number of complaints received from foreign buyers of U.S. grain increased in fiscal year 1985 as compared to the past several years. The complaint system may not reflect the total situation, however. Foreign purchasers are not always inclined to use the system because the Department of Agriculture can do little to help them resolve their disputes with U.S. exporters. The report noted that the Federal Grain Inspection Service has resisted making certain changes in the Official United States Standards for Grain recommended by the General Accounting Office because of a lack of a majority of industry support and its conviction that the standards are "standards of consensus."

Transforming Farm Financial Goals

* 7.52 *

"Financial Stress in Agriculture: Issues and Implications." Proceedings of the symposium sponsored by the American Agricultural Economics Association Task Force on Financial Stress, November 24-25, 1986, Kansas City, Missouri. *Agricultural Finance Review* 47: (Special Issue 1987).

Thirteen papers by agricultural economists associated with universities and agencies of the federal government examine recent experiences on the nature and extent of the farm financial problems, and evaluate private and public responses to farm financial stress. There was widespread agreement that the farm credit crisis is a long-term adjustment to secular trends calling for a massive downsizing of the industry. The participants discussed a wide range of problems and possible adjustments, including a broad but specific selection of needed research projects.

*** 7.53 ***

Barry, Peter J., and others. *Financial Stress in Agriculture: Policy and Financial Consequences for Farmers.* AE-4621. Southern Regional Research Project S-180. Urbana: Department of Agricultural Economics, University of Illinois, 1986.

In a broad interstate study, several economists examine the possibilities of successful refinancing of samples of farms in various degrees of financial difficulty. The research suggests that the best chances for needed and successful refinancing will occur among those farms in the intermediate leveraged range (41 to 70 percent debt-to-asset ratio). The study concludes, however, that "the exceptions among the states indicate the need for a comprehensive financial analysis to demonstrate increasing viability in a public program."

*** 7.54 ***

Barry, Peter J., ed. *The Farm Credit Crisis: Policy Options and Consequences.* AE-4612. Urbana: College of Agriculture Cooperative Extension Service, University of Illinois, 1986.

Seventeen economists from nine state universities and the federal government explored various aspects of the farm credit crisis of the mid 1980s and discussed a range of policy options.

*** 7.55 ***

Buccola, Steven T. "Testing for Nonnormality in Farm Returns." *American Journal of Agricultural Economics* 68:2 (May 1986):334-343.

Cash returns from farming are non-normally distributed under a wide range of joint price-yield distributions. Buccola shows that adequate testing for such non-normality requires use of proper whitening procedures and appropriate statistics. With certain tests and sample sizes, a false imputation of normality often will be made, the usual but positive correlation between skewness and kurtosis reduces the likelihood of associated decision errors. Buccola illustrated these results with data for irrigated alfalfa and dryland wheat.

*** 7.56 ***

Dunford, Richard W. "Farming the Tax Code: Preferences Lower Taxes on Farming, but Is Farming Sector Helped?" *Choices: The Magazine of Food, Farm, and Resource Issues* (Third Quarter 1986):19-23.

The two most important tax preferences exclusively applicable to farming are cash accounting and the deductibility of certain capital expenditures. Dunford concludes that, although these preferences have substantially benefited some farmers directly through lowering their income tax liability, they do not necessarily help the farm sector. "A wide variety of taxpayers qualify for the farm tax preferences. Hence, many non-farm individuals receive some of the resulting tax benefits....there are a few limitations on the entry of resources into farming....lower farm prices as a result of indirect supply effects may have more than offset the direct benefits of farm tax preferences....tax preferences necessitate higher tax rates to raise a given amount of tax revenues....farmers who utilize few of the tax preferences and have a positive taxable income probably have a greater tax liability than they would have if tax rates were lower."

*** 7.57 ***

LeBlanc, Michael, and James Hrubovcak. "The Effects of Tax Policy on Aggregate Agricultural Investment." *American Journal of Agricultural Economics* 68:4 (November 1986):767-772.

The authors concluded "that tax policies are effective in promoting agricultural investment. Nearly 20 percent of net investment in agricultural equipment during the period 1956 through 1978 is attributed to tax policy. From 1956 to 1978, tax policy has stimulated, in real dollars, over $5 billion in net investment in equipment and in excess of $1 billion in structure."

*** 7.58 ***

Reid, Donald W., and Garnett L. Bradford. "A Farm Firm Model of Machinery Investment Decisions." *American Journal of Agricultural Economics* 69:1 (February 1987):64-77.

Reid and Bradford present a multiperiod mixed integer programming (MMIP) model of optimal machinery decisions. They conceptualized infinite horizon valuation models of replacement and other investment situations in the context of a finite programming model. They used dual properties of MMIP model to identify and value opportunity costs involved in investment decisions of farm machinery. They concluded that the interdependent nature of investment and production relationships necessary for solving these values emphasize the importance of a holistic firm perspective in analyzing farm machinery investment strategies.

*** 7.59 ***

Saulnier, R.J., Harold G. Halcrow, and Neil H. Jacoby. *Federal Lending and Loan Insurance.* A study by the National Bureau of Economic Research, New York. Princeton, NJ: Princeton University Press, 1958. LC 57-5489.

This is the first, comprehensive study of federal lending and loan insurance contained in a single volume. It gives the history of the development of federal programs to the mid 1960s, and analyzes their organization, costs, and impacts on the U.S. economy. Separate sections on agriculture give information on federal lending and loan insurance programs for farms and agribusiness firms, including special programs for farm cooperatives.

*** 7.60 ***
Sexton, Richard J., and Terri Erickson Sexton. "Taxing Co-ops: Current Treatment Is Fair, but Not for Reasons Given by Co-op Leaders," and "Taxing Co-ops: Part II, Current Treatment Doesn't Harm the Economy." *Choices: The Magazine of Food, Farm, and Resource Issues* (Second Quarter 1986):21-25 and (Third Quarter 1986):16-18.

The main regulation governing taxation of farm cooperatives, in subchapter T of the federal tax code, allows co-ops to deduct distributions of patronage refunds to members before calculating their corporate income tax. The refunds must be based on business done by patrons with the co-op, and the patron is liable for tax on the refund, just as any other income. Co-op income that is not paid out to patrons is taxable at ordinary corporate rates. The Sextons argue that this is fair since "the tax-free income transfer from co-op to patron is no different from the right of vertically integrated corporations to have income from vesicle subsidiaries be taxable only to the present company." In Part II, they argue that the growth of seldomly restricted co-op membership has improved markets. "Co-ops seldom have sufficient market power to control supply; and, in most cases, co-ops act to correct market failure, not to cause it."

*** 7.61 ***
Sisson, Charles Adair. *Tax Burdens in American Agriculture: An Intersectoral Comparison*. Ames: Iowa State University Press, 1982. ISBN 0813816807.

The purpose of Sisson's study was to "examine farm-nonfarm tax burdens in order to determine whether farmers have a substantial tax advantage over the general population" and to determine "if farmers received tax preferences, who benefits most?" His conclusion is that farmers, and larger farmers in particular, have significantly lower tax burdens compared to other taxpayers. Sisson wrote primarily for professional research workers and the book has much to offer this audience.

*** 7.62 ***
Taylor, C. Robert. "Risk Aversion Versus Expected Profit Maximization with a Progressive Income Tax." *American Journal of Agricultural Economics* 68:1 (February 1986):137-143.

Taylor introduces the terms "apparent," or "pseudo," risk aversion to represent those situations where "real" risk aversion is falsely attributed because random variables are arguments in a decision maker's objective function. Apparent risk aversion can occur when "outside" variables enter the objective function. The progressive income tax structure is highlighted as a neglected but important source of apparent risk aversion. Taylor concludes that "Such sources of apparent risk aversion should be recognized, otherwise an incorrect level of risk aversion might be claimed, or risk aversion might be claimed when risk neutrality prevails."

*** 7.63 ***
Young, Renna P., and Peter J. Barry. "Holding Financial Assets as a Risk Response: A Portfolio Analysis of Illinois Grain Farms." *North Central Journal of Agricultural Economics* 9:1 (January 1987):77-84.

Young and Barry explore the possible gains in risk efficiency for the total farm unit by formulating farm portfolios with different proportions of farm and financial assets. The results of a risk programming model for a representative Illinois cash grain farm indicate that low correlations between returns on farm assets and financial assets could reduce the relative variability of the farm's rates of return on assets by 15 to 25 percent, compared to holding farm assets alone. They conclude that "Extensions of the analysis should account for the effects of asset liquidity, taxation, tenure position, and capital structure."

Transforming Soil Conservation Programs

*** 7.64 ***
Batie, Sandra S., and Alyson G. Sappington. "Cross-Compliance as a Soil Conservation Strategy: A Case Study." *American Journal of Agricultural Economics* 86:4 (November 1986):880-885.

The authors estimated the financial effects of a hypothetical cross-compliance program for seventy-six farmers in Gibson County, Tennessee. Several federal program benefits were compared to the least-cost method of achieving one of four different soil displacement limits to determine if the costs of voluntarily participating in a cross-compliance program were greater than the benefits. They concluded that "If cost-sharing were available, farmers would have a positive incentive to cross comply to 5 tons per acre per year on 57.3 percent of all fields. This figure rises to 90.9 percent compliance at 20 tons per acre per year."

*** 7.65 ***
Boggess, W.G., and E.O. Heady. "A Sector Analysis

of Alternative Income Support and Soil Conservation Policies." *American Journal of Agricultural Economics* 63:4 (November 1981):618-628.

Since 1933 Congress has attempted to legislate both higher farm incomes and soil conservation. Boggess and Heady used a national, interregional, demand-endogenous, separable programming model to analyze the potential of alternative policies to achieve simultaneously the dual goals of increased farm income and reduced soil erosion. They concluded "that a conservation-oriented land retirement policy can be designed to achieve an increase in net farm income equivalent to a traditional general land retirement policy, while simultaneously achieving significant reductions in gross soil erosion, chemical input use, and direct government program costs."

*** 7.66 ***

Bosselman, Fred P., and David Callies. *The Quiet Revolution in Land-Use Control*. Washington, DC: Government Printing Office, 1972.

There is a continuing revolution in the United States relating to the goals in shifting land from one use to another, in controlling the investments in land development and conservation, and in the rules relating to ownership and tenure, transfer and inheritance, and certain related land reforms. Local ordinances have often proved inadequate to deal with a wide variety of land-use problems that are required, state, or national in scope. Control of land use is primarily a legal function of the states. Many new state laws have been enacted; and there are recurring movements at the national level to establish more comprehensive controls over land use and conservation programs.

*** 7.67 ***

Braden, John B. "Some Emerging Rights in Agricultural Land." *American Journal of Agricultural Economics* 64:1 (February 1982):19-27.

The rights to use land for agricultural purposes have become increasingly complex, following new policies for soil conservation. When distinguishing between ownership and exchange rights, little change is evident in ownership rights. The added complexity has come mainly in rules governing exchanges. Braden concludes that "Farmers are confronted increasingly with rules which allow specific rights to be limited by government with compensation. These rules retain the flexibility of individual ownership while reflecting a growing awareness that the general welfare depends on wise use of agriculture land."

*** 7.68 ***

Brown, Keith C., and Deborah J. Brown. "Heterogenous Expectations and Farmland Prices." *American Journal of Agricultural Economics* 66:2 (May 1984):164-169.

Heterogenous expectations for the future affect potential farmland buyers. The authors concluded that "It is optimal for each seller to have a reservation price in excess of the value he attaches to the future stream of income attributable to owning the land if he thinks that some potential buyers may be more optimistic than he." A formula for the optimal reservation price is presented, and a numerical illustration is shown. Using this formula with Corn Belt and Lake State data, the authors made an extremely preliminary empirical test of the importance of optimists in determining land prices for 1968-81.

*** 7.69 ***

Burt, Oscar R. "Farm Level Economics of Soil Conservation in the Palouse Area of the Northwest." *American Journal of Agricultural Economics* 63:1 (February 1981):83-92.

Burt applied control theory to the farm-level economics of soil conservation in a model using depth of topsoil and percentage of organic matter therein as the two state variables. He tested an approximately optimal decision rule against the optimal rule and found the optimal decision rule to be excellent; errors in the decision rule were less than 1 percent within the region. He concluded that "intensive wheat production under modern farming practices and heavy fertilization is the most economic cropping system in both the short and long run in the Palouse Area except under low wheat prices."

*** 7.70 ***

Calef, Wesley. *Private Grazing and Public Lands, Studies of the Local Management of the Taylor Grazing Act*. Chicago: The University of Chicago Press, 1960. LC 60-15936.

There is an inherent conflict between the interests of ranchers in using the public domain for grazing, primarily in the western states, and the interests of the general public in conserving soil and water and preserving wildlife and national forest land. After many years of growing conflict, the Taylor Grazing Act of 1934 was passed to reorganize the use and control of the 140 million acres of "vacant, unappropriated, and unreserved lands of...the public domain." Calef recognizes the continuing difficulty of reconciling conflicting interests. Although research for solutions that are socially optimum must still go on, and continuing education must play a major role, the institutions of government are required to manage the inherent conflicts. By the mid 1950s some 160 million acres of public grazing land were brought under control of the Division of Range Management, Bureau of Soil

Management, Department of the Interior. This land, added to the 160 million acres of national forest land under the U.S. Forest Service on which some grazing is permitted, constitutes the bulk of federally owned land used for food production in the 48 states. In addition, the federal government still owns nearly 98 percent of the 365 million acres in the state of Alaska.

*** 7.71 ***

Castle, Emery N., and Irving Hoch. "Farm Real Estate Price Components." *American Journal of Agricultural Economics* 64:1 (February 1982):8-18.

The research on which Castle and Hoch reported demonstrated that farm real estate price involves important components in addition to the capitalized value of rent for the services of land and buildings in farm production. They developed an expectations model for the farm real estate market, and compared predictions from the expectations model with farm real estate prices for the 1920-78 period. They concluded that "Capitalized rent explains only about half of real estate values both in the 1970s and over the longer 1920-78 period. The remainder can be explained by the capitalization of capital gains, including real gains or losses from price level changes."

*** 7.72 ***

Clawson, Marion. *The Land System of the United States*. Lincoln: University of Nebraska Press, 1968. LC 68-10250.

Clawson discusses the reasons that the land system was established and how it operates. He treats the subject with sympathy and understanding based on his many years of service as a land economist with the federal government, various universities, and a private foundation.

*** 7.73 ***

Crosson, Pierre. "Soil Conservation: It's Not the Farmers Who Are Most Affected by Erosion." *Choices: The Magazine of Food, Farm, and Resource Issues* (Premiere Edition 1986):33-38.

Research shows that erosion is increasing, raising serious questions about erosion policies and programs. In summarizing some of the research, Crosson concludes that "the present annualized costs of cropland erosion on soil productivity are about \$1.7-\$1.8 billion." Because of technological advance, "Continuation of present erosion rates would not likely impose significantly higher costs of food and fiber on future generations." However, as he notes, "A Study by the Conservation Foundation indicates that off-farm erosion damage currently costs the nation \$3.4 billion - \$13.0 billion annually, with the 'best guess' estimate being \$6.1 billion." He concludes that "Soil conservation policy as it has been known since its inception in the 1930s is on the verge of a fundamental transformation."

*** 7.74 ***

Foss, Phillip O. *Politics and Grass, the Administration of Grazing on the Public Domain*. Seattle: University of Washington Press, 1960. LC 60-11822.

Foss discusses the inherent role of politics in contending with conflicting interests in management of the public domain. He shows that there is a need for public management and that the issue with which this management must deal will continue.

*** 7.75 ***

Foster, William E., Linda S. Calvin, Grace M. Johns, and Patricia Rottschaefer. "Distributional Welfare Implications of an Irrigation Water Subsidy." *American Journal of Agricultural Economics* 68:4 (November 1986):778-786.

The authors analyzed the distributional welfare implications of a subsidy for irrigation water for California rice producers. A more general equilibrium approach than that used in previous studies was taken in order to determine the effects of the subsidy on consumers, subsidized producers, and unsubsidized producers. The two important policy conclusions of the authors are "that unsubsidized producers bear part of the cost of a subsidy through lower prices, and that consumers (taxpayers) may gain by sponsoring increased production through a selective subsidy."

*** 7.76 ***

Gunterman, Karl, M.T. Lee, A.S. Narayanan, and E.R. Swanson. *Soil Loss from Illinois Farms, Economic Analysis of Productivity Loss and Sediment Damage*. IIEQ Document, no. 74-62. Chicago: Illinois Institute for Environmental Quality, 1974.

The authors analyze the economic effects of soil erosion and sediment damage on six watersheds in Illinois, selected for study to cover a wide range of soil, slope, and erosion conditions. Although erosion and sediment damage is highly correlated with the intensity of cultivation and cropping, the on-site soil losses, expressed as a reduction in net rents per acre, were typically very low. In most instances, potential soil losses could be ignored by farmers in planning their crop rotation, tillage system, and conservation practices. The off-site sediment damage might be much higher, running from a few cents per cropped acre to over \$3 per acre for the most intensive cultivation.

*** 7.77 ***

Halcrow, Harold G., Earl O. Heady, and Melvin L. Cotner, eds. *Soil Conservation: Policies, Institutions, and Incentives*. Ankeny, IA: Published for

North Central Research Committee 111, Natural Resource Use and Environmental Policy, by the Soil Conservation Society of America, 1982. ISBN 0935734-06-6. LC 82-699.

A symposium, "Soil Conservation Policies, Institutions, and Incentives," was held May 19-21, 1981, at Illinois State Park, Zion, Illinois. The editors suggest that it provides a "perspective for conservation efforts, especially the actions by the public sector." The role of the private sector and individual participants in conservation decisions is examined. Another section deals with the reexamination of the conceptual base for public and private action to conserve soil. Finally, alternative strategies to achieve conservation are evaluated. The entire discussion is intended to interest managers, administrators, and legislators who plan and implement activities that contribute to informed decisions about the use and management of soil and water resources.

*** 7.78 ***

Hardin, Charles M. *The Politics of Agriculture, Soil Conservation and the Struggle for Power in Rural America.* Glencoe, IL: The Free Press, 1952. LC 52-8160.

Hardin offers a penetrating study of the political forces surrounding the development of soil conservation programs. The political struggle involves the general farm organizations as well as many of the special-interest groups that attempt to influence farm policy. Hardin shows how the various interests work to influence policy and why such groups are recognized in the political process by which soil conservation policy is made.

*** 7.79 ***

Hibbard, Benjamin H. *A History of Public Land Policies.* New York: The Macmillan Press Company, 1924. LC 24-28692.

Hibbard's book was the first general history of the public land policies and was widely used as a text in colleges and universities. In many respects, it was a classic of its time and set a high standard of scholarship for students of land policy.

*** 7.80 ***

Miranowski, John A., and Katherine H. Reichelderfer. "Resource Conservation Programs in the Farm Policy Area." In *Agricultural-Food Policy Review: Commodity Program Perspectives.* Agricultural Economic Report, no. 530. Washington, DC: Economic Research Service, U.S. Department of Agriculture, 1985.

The primary purpose of soil and water conservation programs is to maintain our agricultural productive capacity over time. The authors state "to assess the role of these resource conservation programs in developing future legislation, an understanding is needed of the soil and water resource problems, the rationale for conservation spending, the historical evolution of the programs, and the effectiveness of current programs." This information can be used to achieve greater efficiency in future programs and realize more consistency between conservation and commodity programs.

*** 7.81 ***

Morgan, Robert J. *Governing Soil Conservation: Thirty Years of the New Decentralization.* Baltimore: The Johns Hopkins Press, published for Resources for the Future, 1965. LC 65-27670.

Morgan has written a comprehensive study of soil conservation programs, mainly under the administration of the Soil Conservation Service from passage of the Soil Conservation Act of 1935 to the mid 1960s. This act, as amended by the Soil Conservation and Domestic Allotment Act of 1936 and subsequent legislation, became the parent legislation for soil conservation programs in succeeding years.

*** 7.82 ***

Phipps, Tim T. "Land Prices and Farm-Based Returns." *American Journal of Agricultural Economics* 66:4 (November 1984):422-431.

Phipps developed the theoretical and empirical relationship between farm-based residual returns, opportunity costs of farmland, and farmland prices. He tested temporal hypotheses concerning the source of land price movements using a variant of Granger causality and found that farmland prices are unidirectionally "caused" by residual farm-based returns. Phipps concluded that "The findings support the hypothesis that farmland prices are determined mainly within the farm sector and lend credence to the use of extrapolative expectations processes in structural farmland price models."

*** 7.83 ***

Pope, C. Arden, III. "Agricultural Productive and Consumptive Use Components of Rural Land Values in Texas." *American Journal of Agricultural Economics* 67:1 (February 1985):81-86.

Pope found that "Consumptive demand applies significant upward pressure on rural land values and plays an important role in determining farm and ranch structure in Texas." He concluded that "Population density, proximity to major metropolitan centers, quality of deer hunting, and aesthetic differences across the state explain the majority of the differences in rural land values. On the average, only about 22 percent of the total market value of rural land in Texas can be statistically explained by its productive value."

*** 7.84 ***

Salter, Leonard A., Jr. *A Critical Review of Research in Land Economics.* Minneapolis: The University of Minnesota Press, 1948. LC 48-2149.

The author provides a penetrating analysis of the concepts, accomplishments, and weaknesses of research in land economics, based on a Ph.D. dissertation presented to the University of Wisconsin, published posthumously. The book has been a basic reference for students of land economics and many of its conclusions are of timeless importance.

*** 7.85 ***

Swanson, Earl R., A.S. Narayanan, M.T. Lee, Karl Gunterman, and W.D. Seitz. *Economic Analysis of Erosion and Sedimentation.* AERR 126, 127, 128, 130, 131, and 135. Urbana: Department of Agricultural Economics, University of Illinois, 1974-1975. 6 vols.

Each of the six publications provides an economic analysis of soil erosion and sediment damage on a selected watershed in Illinois. Erosion and sediment damage are highly correlated with the intensity of cultivation and cropping. Continuous corn and rotations of corn and soybeans, which are the most profitable combination over much of the cash-crop areas of Illinois, as well as other states in the cornbelt, have significantly higher rates of erosion and sediment damage than rotations that also include wheat and meadow crops. The on-site soil losses in the six Illinois watersheds, expressed as a reduction in net land rents per acre, were very seldom a significant cost to the farmers, even though some of the watersheds had significant off-site damages from erosion.

*** 7.86 ***

United States. Department of Agriculture. Committee on Land Use. *Perspectives on Prime Lands.* Washington, DC: U.S. Government Printing Office, 1975.

Background papers dealing with the problem of preserving the use of highly productive land for production of food and other farm products have been collected in this volume. The papers present a broad perspective on the basic issues involved in the competition among alternative uses for this land and provide a useful basis for public policy.

*** 7.87 ***

United States. Economic Research Service. *Sodbusting: Land Use Change and Farm Programs.* Agricultural Economic Report, no. 536. Washington, DC: U.S. Department of Agriculture, 1985.

In the United States between 1979 and 1981, farmers converted about 11.1 million acres of land to cropland uses, but only 1.9 million acres were both highly erodible and planted to program crops. Although such conversion has been occurring in all regions, the Great Plains is the area of greatest concern about the effects of such conversion. The study concludes that "Analysis of costs and returns indicates that farm programs do provide an incentive to convert highly erodible land to cropland. Participation in price support and subsidized loan programs would have made net returns on 384,000 acres of highly erodible land profitable in 1982. Proposed legislation would remove such incentives, but the proposed system for identifying highly erodible land does not precisely identify new cropland with high potential for excessive erosion."

*** 7.88 ***

Webb, Shwu-Eng H., Clayton W. Ogg, and Wen-Yuan Huang. *Idling Erodible Cropland: Impacts on Production, Prices, and Government Costs.* Agricultural Economic Report, no. 550. Washington, DC: Economic Research Service, U.S. Department of Agriculture, 1986.

To identify erodible and fragile land, the authors developed land group criteria that link productivity with potential soil erodibility. About 32 million acres of cropland in the United States were identified as highly erodible and fragile. They estimated the impact of idling those acres on production and prices for seven major crops (corn, soybeans, wheat, sorghum, oats, barley, and cotton) under assumptions generally consistent with recent farm legislation. They concluded that "A government program to put erodible land into a conservation reserve would reduce soil erosion and complement the goals of commodity programs by supporting crop prices and reducing government deficiency and storage payments."

CHAPTER 8

EDUCATIONAL PROGRAMS FOR IMPLEMENTING CHANGE

"An investment in knowledge pays the best interest."
Benjamin Franklin, *Poor Richard's Almanac.*

"...if breakthroughs in biotechnology prove to be as explosive as sometimes predicted...the land grant university may face a searching of its corporate soul as to whether it will facilitate a tailoring of that technology to a decentralized agriculture, or will let it be the agent of shift into an agriculture of conglomerate corporations."
Harold F. Breimyer[1]

The modernization of farming may be partly credited to investments in education, skills and health of people in farming, agribusiness, and industry related to agriculture. Farmers and farming operations will continue to face adjustments perhaps even more revolutionary than in the past.

The modernization of farming may be partly credited to investments in education, skills and health of people in farming, agribusiness, and industry related to agriculture. These investments result in rising rates of productivity of people and in declining requirements for certain amounts of agricultural output.

Changes in farming will continue. Farmers and farming operations will continue to face adjustments perhaps even more revolutionary than in the past. The changes resulting from the development of biotechnology and its impacts upon farming operations and individual farmers could bring about a revolution in farming as significant as tractors and electricity did in the first half of the twentieth century.

Historical Background

Much of the success measured by rising productivity in American farming is credited to the creation of its agricultural research and extension system. For well over a century, the federal and state governments have invested substantial sums of money in agricultural research and extension. In the mid-1980s, the expenditures were about $3 billion annually.[2]

The cooperative system of agricultural research and the extension of that research into practical fields of application, as carried out by the Department of Agriculture and the land grant universities is one of the oldest farm-related activities of our federal government. It was founded on the belief that the applications of scientific methods to the problems of farming would enhance the welfare of farmers and rural citizens and improve the food supply for all.

Several important events have helped make the agricultural research and extension system an integral and longstanding part of U.S. agricultural policy.

In 1796 President George Washington recommended the establishment of an agricultural branch of the national government. In 1841 the president of Norwich University proposed to Congress that it appropriate funds from land sales to be distributed to the states for establishing institutions to teach agriculture. During the 1840s and 1850s, state legislatures, farm leaders, the editors of agricultural periodicals, and farm organizations, particularly the United States Agricultural Society, urged Congress to act on both of these proposals.[3]

The milestone legislation that established an agricultural research and education policy for the United States came in 1862. President Abraham Lincoln signed the legislation that established the Department of Agriculture and the Morrill Act, which provided for land grants to states to establish agricultural and engineering colleges. Later, the Hatch Act in 1887 provided for federal funds to help the land grant universities conduct agricultural research and the Smith Lever Act in 1914 established extension work for farmers and farm families.[4]

In 1890 Congress authorized additional land grant colleges in 17 southern states to serve primarily black students. These later became universities and have been integrated for students of all races. They have received additional funding for research and extension programs in recent years and have special advantages to study problems unique to small farmers.

So the land grant universities located in every state differ from other public and private universities. Not only do they teach resident agricultural students, but their faculty engages in research, supported in part from federal funds appropriated through the Hatch Act, and in extension educational programs supported in part through funds authorized under the Smith Lever Act for farmers, homemakers, and youth in locations across the state.

New Technology and the Education Issues Facing American Farmers

In recent years, the land grant universities have been criticized for directing their work too narrowly to plants, animals, soils, and farm prices to the neglect of the problems of small farmers, rural communities, and consumers.

The Office of Technology Assessment (OTA), a unit established by Congress to study technology issues, has raised some crucial questions about agricultural research and extension policy, and the consequences for the future structure of American agriculture.[5]

Who Profits from Technology Change?

The first farmers to adopt new technologies are considered the immediate beneficiaries. Their cost per unit of production are lowered and their profits are thus increased. The business firms that supply the products of new technology also benefit. Higher profits encourage the adopting farmers to expand output--even to the extent of increasing the size of their farm operation. Farmers who adopt the new technology later realize less profits as prices may decline from larger production.

Three lessons may be derived: Those farmers who are most aggressive in effectively adopting and applying new technologies are the most likely to survive. Their size or scale of operation thereby influences the structure of agriculture. Likewise, structure is affected to the extent that research discoveries or extension programs favor farm operations of a certain scale. Not only is research and extension vital for individual farmers, it is also vital to maintaining the competitive position and comparative advantage of U.S. farmers in international trade.

Learning to Use New Technology

A major question is how much science and education should come from privately funded sources and how much from public sources. In recent years, private industry has played a role in education for farmers. For most agribusiness firms, this role relates to their efforts to promote the products and services that they market. The education value of these promotional activities related more to alerting farmers to the availability of new products than to evaluating objectively the performance of those products.

Generally, the Extension service has played the role of evaluation since its involvement is more efficient and less costly for the farmer. Under the Cooperative Extension Service system, federal, state, and local government funds are used to provide the traditional educational programs serving farmers and other citizens.

Under the concept of new "Federalism" and efforts to reduce the federal deficit, federal funds for extension programs are likely to remain steady or decline in years to come. State and local government units will be asked to pay a higher share of the cost. User fees for programs, printed materials, computer software, and other services are becoming more and more prevalent.

Education for Farmers of the Future

Ever since land grant universities were established, colleges of agriculture have set the standards for higher education in agriculture. Farm youth and those interested in farm and agribusiness careers usually enrolled in their state-supported college of agriculture.

The future agricultural role of the land grant universities is becoming more uncertain. In many of the major land grant universities, the number of students, the number of faculty, and funds devoted to agriculture are substantially less than those devoted to the arts and sciences. Many agricultural officials fear that declining enrollments will lead to declining support for teaching, research, and extension.

The crucial question then becomes whether the USDA-Land Grant university system must be redesigned to meet the educational needs for farmers of the future. The era of biotechnology research presents important challenges for extension and research directors in the land grant universities. How can they be sure that extension educational programs will bridge the gap from basic research to applied demonstrations and educational information that a farmer can use and be competitive in a farming system growing increasingly complex? With research scientists conducting more basic research, many of those directing extension programs may also have to do applied research to support those programs.

In an age of information, important questions must be raised about the formal education, as well as the continuing education, for farmers in the future. With more and more farms using highly specialized technology and marketing from $300,000 to $1,000,000 of products a year, the need for business management, accounting, and financial strategies may lead young people planning careers in agricultural production to seek formal education in colleges of business as well as in colleges of agriculture. The M.B.A. degree for farm operators may become just as useful as for other business owners and executives.

The Farmer as Human Capital

The term human capital became prominent in the economics literature in the early 1960s when it was used to describe the processes through which people increase their future productivity. Education is probably the most common way for forming human capital. The public recognizes a social advantage that accrues to an educated population and public policy has established public elementary and secondary schools. Beyond that minimum each individual must bear more of the cost of acquiring additional knowledge or skill.

Each generation of farmers, along with the rest of the population, has acquired more formal education than their parents. In a survey of farmers taken in 17 states in 1984, from 33 to 61 percent of the respondents reported formal education beyond high school.[6]

The beginning farmers who are successful generally have made substantial investments in their own human capital in the form of education and training. They often have additional help from family, relatives, or friends. Some also will take advantage of more educational opportunities after entering farming, thus increasing their skills more than others. As some farmers become more efficient and productive, those who have not increased their human and other forms of capital tend to lose relatively.

From the perspective of economic analysis, more education and investment in human capital in agriculture will increase the opportunity cost--the income available to educated farmers in some off-farm occupation--assuming there are attractive employment opportunities in the off-farm sector of the economy. The well-educated person who has invested for a farming career could also find attractive career opportunities outside of direct farm production.

Education for Farmers in Distress

When farmers and farm families across the nation faced financial distress in the mid-1980s, new forms of educational effort emerged to help them. Several state extension services set up farm crisis hotlines. These enabled financially stressed farmers to get farm management and financial counselling through confidential appointments with experienced counsellors.

Other organizations established farm crisis telephone lines to provide varied forms of assistance. Churches and social service agencies frequently established support groups where farm couples with similar problems could meet and discuss their mutual concerns and help each other cope with the crises they were facing.

Because of the private and personal nature of financial problems, the most effective education for farmers in financial crisis is often on a one-to-one personal counselling basis. Although some farmers might be enabled to continue farming with such counselling, the most useful part of such programs was often to guide families through serious mental stress, which, without such help, could lead to suicides and other serious consequences.

State Efforts to Help Farmers under Financial Stress

State governments can do little to change the root causes of farm financial problems. They do not control macroeconomic policy or have the resources to raise farm income. Yet, beginning in the late 1970s, state government policies began to develop the capacity to influence the number of family farms that could survive, help displaced farmers move into new occupations, and encourage economic development in rural areas. A wide range of programs and services have emerged under the auspices of different states.

The most frequent assistance has been special telephone "hotlines," a special number, usually toll-free, that distressed farmers can call for counselling and referral to sources of help, with assurance that such inquiries will be confidential.

Another significant area of assistance programs has been credit and finance. Some states have established special credit programs to enable farmers to borrow money at lower rates of interest, reduce interest costs on current loans, or defer interest until the farm income situation improves. Some states have established mediation procedures to help farmers and lenders resolve their problems or have placed moratoria on farm foreclosures.

Educational Programs for Those Who Leave Farming

Traditionally, farmers who earned less from farming than they could from non-farm labor migrated to non-farm employment. By doing so they were investing in their own human capital--using their ability to turn skills and knowledge into greater income.

With rapid technical change, as occurred in the United States from 1940 to 1980, the need to transfer labor out of agriculture tends to be even greater. Without this movement off-farms, the surplus of labor would mean lower incomes for those who remain in farming. By the mid-1980s, the major part of surplus labor had moved off-farms. Yet rapid change and economic distress continued to force some farmers into non-farm occupations.

The opportunity to migrate successfully is closely linked to the number of employment opportunities, the individual's range of abilities, skills and encumbrances, and the individual's degree of willingness to move to take another job.

Regardless of causes, time, or name, the major problem has persisted through present and past crises: there are too many resources in agriculture and the industry cannot adjust until some of these resources are removed.

Labor can be removed by investing in training or migration so that present farmers can carry out their productive lives in nonagricultural employment with dependence on salaries and wages rather than unpredictable farming returns. These farmers may require selective retraining. Or participation by entire communities may be needed to help farm operators and their families find new sources of income.

CHALLENGES FOR THE FUTURE

Appropriate Education and Training

In the new era of advanced biotechnology in agriculture, what will be the appropriate education and training for the farm operators of the future? In the early part of the twentieth century, Congress passed the Smith Lever Act to further Extension work among farmers, homemakers, and farm youth. In 1917, the Smith-Hughes Act provided for federal support for vocational agriculture in high schools. These Acts served the farming community well for many years.

After World War II, the community college system in many states offered advanced training in agriculture beyond high school. A higher percentage of farm youth went on to four-year college programs in agriculture and other fields. With pressures for more science and basic subjects for college entrance, high school vocational agriculture courses received less credit for college entrance requirements and enrollment declined.

Development of human capital to operate the farms of the 1990s and the 21st century will present a major challenge to educators and those vitally interested in maintaining a system of dispersed family operated farms. The elementary, secondary, community college, and four-year college programs will all need a reappraisal and reorganization to meet the needs of the operators and managers of high technologically-equipped farms in the future.

Goals for Education in Agriculture

A major concern among those who strongly advocate a family system of agriculture is that new technology and increasing capital requirements will make family operations difficult if not impossible to survive. A crucial policy issue concerning education for farmers is whether education can provide the means to preserve a system of family farms.

Another closely related question is whether

education should have as one of its purposes to preserve family farms. Or should the free enterprise system prevail and only those who can survive in a free competitive market be allowed to continue in farming?

Research and Continuing Education

Rapidly changing technology and the rapid advancements in information development and transmission will require continuous updating for progressive and successful farmers in the future. Public policy has encouraged private development of plant varieties that could be patented, new agricultural chemicals and herbicides by competing private manufacturers, and private research development on various biotechnology projects.

The role of the public and private sectors in providing farmers with new information and technology will affect the future structure of agriculture--the numbers, sizes, and control of the farming units. How new technology becomes available, its form, cost, and adaptability to various sizes of farming operations will determine whether the single family farm can survive or if future farming units will be mainly multi-family and corporate forms of business organization. The investment in human capital by the farm operators and their capability to use new technology in an economically efficient manner will become an ever greater determining factor for their survival in farming.

License to Use Technology

Emphasis on pollution control and protection of the environment has brought the introduction of special certification for farmers to use certain pesticides. This certification has been carried out through training programs conducted by the Extension Service, sometimes in cooperation with state departments of agriculture. As new and potentially dangerous agricultural chemicals and other biotechnology are introduced, a continued emphasis on proper and safe use will be part of the certification procedure to permit their use.

The crucial issue will be what kinds of educational programs will be required, who will conduct them, what formal educational programs will be required for farmers to pass these tests, and whether such licensing could be used to restrict use of technology or limit the number of farmers who can use it.

Information Transmission

By the mid-1980s, the computer age had arrived on American farms. The most progressive and innovative farmers were using computers to keep farm business records, analyze the most efficient use of farming inputs, gather market information and analysis, and communicate with others by means of electronic mail and bulletin boards.

The most effective use of the new information technology requires specific investment in human capital and continued investment time to keep abreast with equipment and software programs available. Farmers of the future will include information technology as part of their formal and continuing education. The farm business will need to be large enough to support the investment in human capital and equipment.

Conclusions

Successful farmers in the future will require more formal education, a broader education that includes both production technology and business management, and a continuing education program similar to those now used by various professions.

Agricultural research will be conducted by both private companies and public institutions. Public policies will guide the methods by which new research developments are applied and disseminated to farms and related agricultural businesses. The distribution of new agricultural technology will be watched closely by environmental and regulatory agencies to make sure that such new introductions will not have adverse effects.

For farmers who lack adequate capital or land to develop a financially viable operation, educational programs to assist in their movement to off-farm employment will be developed and used more frequently.

Endnotes

1. Harold F. Breimyer. *Economic Issues*, no. 100. Madison: Department of Agricultural Economics, College of Agriculture and Life Sciences, University of Wisconsin, 1986.

2. Orville G. Bentley. Assistant Secretary of Agriculture for Science and Education, testimony before the House Agriculture subcommittee on Department Organization, Research and Foreign Agriculture, March 27, 1985.

3. Wayne D. Rasmussen, and R. J. Hildreth. "The USDA-Land Grant University System in Transition." *The Farm and Food System in Transition*, no. 16. East Lansing, Michigan State University. Cooperative Extension Service, 1983.

4. Rasmussen and Hildreth. op. cit. p.1.

5. United States. Congress. Office of Technology Assessment. *Technology, Public Policy and the Changing Structure of Agriculture*, OTA-F-285, Washington, DC: U.S. Government Printing Office, 1986. Chapter 12.

6. Harold D. Guither, Bob F. Jones, Marshall A. Martin, Robert G. F. Spitze. *U.S. Farmers' Views on Agricultural and Food Policy*. North Central Regional Extension Publication, no. 227. Urbana: College of Agriculture, University of Illinois, 1984.

Chapter 8: Annotated Bibliography

Historical Background

*** 8.1 ***

Glaser, Lawrence K. *Provisions of the Food Security Act of 1985*. Agriculture Information Bulletin, no. 498. Washington, DC: Economic Research Service, U.S. Department of Agriculture, 1986.

Under the National Agricultural Research, Extension and Teaching Policy Act of 1977, a part of the Food and Agricultural Act of 1977, the U.S. Department of Agriculture and the land grant universities are assigned key roles in agricultural science and education. In amendments passed as part of the Food Security Act of 1985, the secretary of agriculture is charged with responsibility for coordinating efforts of state land grant universities, Extension Services, and USDA agencies in developing a plan for and transfer of new technologies, particularly biotechnology, to the farming community. Small- and medium-sized farms are to be given special emphasis.

*** 8.2 ***

McGranahan, David A., Don C. Hession, Fred K. Hines, and Max F. Jordan. *Social and Economic Characteristics of the Population in Metro and Nonmetro Counties, 1970-80*. Rural Development Research Report, no. 58. Washington, DC: Economic Research Service, U.S. Department of Agriculture 1986.

The measure most commonly used to assess the educational level of a population is the median number of school years completed by people age 25 and over who have generally finished school. Nonmetro areas have overcome their lag in educational attainment, and with a median education of 12.3 years in 1980, were about equal with metro areas' median of 12.6 years. This report documents changes in the economic and social characteristics of both metro and nonmetro residents from 1970-80.

*** 8.3 ***

United States. Department of Agriculture-National Association of State Universities and Land Grant Colleges Committee on the Future of Cooperative Extension. *Extension in the '80s*. Madison: Cooperative Extension Service, University of Wisconsin, 1983.

The Cooperative Extension Service, a unique achievement in American education, is an agency for change, a catalyst for individual and group action with a history of 70 years of public service. This report is a summary of the findings of a special joint study committee. Its charge was to review and restate the roles and responsibilities of each of the partners in Cooperative Extension and to produce a document that will serve as a guide for the future mission.

New Technology and the Education Issues Facing American Farmers

*** 8.4 ***

Prawl, Warren, Roger Medlin, and John Gross. *Adult and Continuing Education through the Cooperative Extension Service*. Columbia, MO: University of Missouri Printing Services, 1984.

The authors seek to bridge the gap between the first 50 years of a conservative rural-oriented extension service and the more responsive, rapidly changing organization of the 1980s. Many of the topics are included in extension education courses taught through the land grant university system. Extension Service's dynamic, localized approach to the solution of the common person's problems has stood the test of time. What is necessary is a revitalization of the existing system through new innovations that focus on accountability and efficiency.

*** 8.5 ***

Ruttan, Vernon W. *Agricultural Research Policy*. Minneapolis: University of Minnesota Press, 1982. ISBN 0816611017, 0816611025 (paperback).

During the 20th century, agriculture has been undergoing a transition from a resource-based sector to a science-based industry. Growth in agricultural output is increasingly based on development of scientific and technical capacity to invest new mechanical, chemical, and biological technologies. The issues related to the strategy and organization of agricultural research have rarely been the subject of scientific inquiry. Ruttan points out that effective research management is the product of a unique combination of experience, insight, skills, and personality.

*** 8.6 ***
United States. Department of Agriculture. *After a Hundred Years.* The 1962 Yearbook of Agriculture. Washington, DC: U.S. Government Printing Office. LC agr62-298.

Rather than an ultimate book on agriculture research, this volume provides an enlightening glimpse into just a few aspects of research generated by the USDA, Agricultural Experiment Stations, and a few private sector scientists. Included are sections on biotechnology, its application in the microbial world, to animals and plants, insects and weeds, human nutrition and food, and forest resources. Final sections discuss the agricultural agent of the future, information centers, and new computer technology along with careers in agriculture.

*** 8.7 ***
United States. Department of Agriculture. *A Time to Choose.* Summary Report on the Structure of Agriculture. Washington, DC: U.S Government Printing Office, 1981.

"Our agriculture today is at a crossroads," the authors declare. Continuing existing policies and programs without change will almost certainly mean that present structural trends will continue. The underlying theme of this publication is that concentration of production is undesirable and that steps must be taken if current directions are to be changed. Secretary of Agriculture Bergland who spearheaded the dialogue concluded that more communication and less confrontation between those with different views is needed to resolve differences and develop public policy on the future structure of agriculture.

Who Profits from Technology Change?

*** 8.8 ***
Busch, Lawrence, J. Lew Silver, William B. Lacy, Charles S. Perry, Mark Lancelle, and Shripad Deo. *The Relationship of Public Agricultural R & D to Selected Changes in the Farm Sector.* A Report to the National Science Foundation. Lexington: Department of Sociology, University of Kentucky, 1984.

The authors have designed this study to demonstrate the effect of publicly supported agricultural research and extension on the commercialization of farming, the concentration of farm production, and change in the composition of the labor force. Early adopters of new technology tend to capture the benefits of adoption while late adopters merely avoid losses that would occur if they retained the old technology. Previous studies of public and or private research and development investment have focused on productivity and have largely ignored distributional effects.

*** 8.9 ***
Eddleman, B. R. "Research and Development for the Future Farm and Food System." *The Farm and Food System in Transition*, no. 24. East Lansing: Cooperative Extension Service. Michigan State University, 1984.

New and more efficient technologies for primary farm production, food manufacturing, and distribution stem from both fundamental and applied research. Private businesses utilize both public knowledge and the results of their own research in developing products and processes. This interface of public and private sectors helps secure an efficient, competitive farm and food system.

*** 8.10 ***
Fishel, Walter L., and Martin Kenney. "Challenge to Studies of Biotechnology Impacts in the Social Sciences." In *Agriculture in a Turbulent World Economy.* Proceedings of the Nineteenth International Conference of Agricultural Economists. Oxford: Institute of Agricultural Economics, Oxford University 1986. ISBN 0-566-05225-3.

"Proper and rapid response to anticipated impacts of biotechnology introductions will be much too important not to provide less than the best possible information that is clearly oriented to the decision- and policy-makers needs," the authors emphasize. Social scientists interested in the consequences of technical change on agriculture and related economic and social infrastructures must begin to analyze the implications of new product introductions and processes resulting from biotechnology. The changes from biotechnology may surpass the effects of the Green Revolution and have different impacts on the economic and social structures of the developed countries.

*** 8.11 ***
Hadwiger, Don F. *The Politics of Agricultural Research.* Lincoln: University of Nebraska Press, 1982. ISBN 0803223226. LC 81-24077//r842.

Hadwiger gives agricultural research high marks for its contributions to the production in the United States of ample supplies of food and fiber. Also "in the United States, food costs less as a percentage of income than elsewhere in the world...". Cost benefit studies show rates of return from all kinds of agricultural research, which run as high as 90 percent per year. Hadwiger also says that funding for agricultural research is inadequate to meet needs and opportunities. He suggests that this situation might be improved.

*** 8.12 ***

Hightower, Jim. *Hard Tomatoes, Hard Times.* Cambridge, MA: Schenkman Publishing Company, 1973. LC 72-97185.

"Although the land grant college complex was created to be the people's university...the system has, in fact, become the sidekick and frequent servant of agriculture's industrialized elite," Hightower declares. This report of the Agribusiness Accountability Project condemned the land grant university complex for its biases in agricultural research and extension programs. The message of the report was that the tax-paid, land grant complex was serving an elite of private, corporate interests in rural America while ignoring those who have the most urgent needs and most legitimate claims for assistance.

*** 8.13 ***

McCalla, Alex F. "The Politics of the U.S. Agricultural Research Establishment: A Short Analysis." *Policy Studies Journal* 6:4 (Summer 1978):479-483.

The reality of the establishment is quite different than what its formal structure would suggest. "It is an amorphous hulk of disjointed and often competitive components, each of which is being tugged and pulled by myriads of special interests," McCalla asserts. The establishment in its totality is largely impervious to political manipulation in the short run. However, specific interests, particularly if they have money in hand, can materially influence micro priorities. The fact that the system is disjointed, unplanned, uncoordinated, competitive, and slow to change may be its greatest strength.

*** 8.14 ***

Rogers, Everett M. *Diffusion of Innovations.* 3d ed. New York: The Free Press, 1983. ISBN 0029266505.

Rogers' pioneer work combines a revision of the theoretical framework of diffusion of innovations, the research evidence supporting this model, new concepts and new theoretical viewpoints, and a continuity of his two previous books on diffusion. Much of the early work on diffusion research applied to agriculture and the process by which new farming practices were adopted. In recent years the concepts have been expanded to apply to developing countries, rural communities, and organizations.

*** 8.15 ***

United States. Congress. Office of Technology Assessment. "Impacts on Agricultural Research and Extension." In *Technology, Public Policy, and the Changing Structure of American Agriculture.* OTA-F-285. Washington, DC: U.S. Government Printing Office, 1986.

OTA sees the agricultural research and extension system as an important contributor to a plentiful and low-cost food and fiber supply as well as to the positive U.S. balance of agricultural trade. For the period 1945-79, technological innovations brought about in part by the system increased farm output 85 percent with no change in the level of farm inputs. But in the new era of biotechnology and information technology, OTA raises some questions about the impact of technical advances on the performance of the research and extension system and about how that performance will ultimately affect the structure of agriculture, the numbers, sizes, and control of American farms. These questions include: Who gains and who loses from the process of technological change on American farms? Is agricultural research and extension treating all sizes of farms the same or does it favor the growth of large industrialized farms? What are the roles of the various parts of the agricultural research and extension system as they deal with technological changes in the biotechnology and information technology era? How is a proper balance to be developed between public and private components of the agricultural research and extension system?

Learning to Use New Technology

*** 8.16 ***

Joint Council on Food and Agricultural Sciences. *Five-Year Plan for the Food and Agricultural Sciences.* A Report to the Secretary of Agriculture. Washington, DC: U.S. Department of Agriculture, 1986.

The Agriculture and Food Act of 1981 requires that the Five-Year Plan for the Food and Agricultural Sciences be updated every two years. The plan provides evidence that the decentralized research, extension, and higher education system can plan together. It also provides a forum for continued evaluation of goals and objectives, a standard for evaluating progress, a planning aid for decision makers, and an accounting of human resource allocations and projections of anticipated change and needs.

*** 8.17 ***

United States. Congress. General Accounting Office. *Biotechnology: The U.S. Department of Agriculture's Biotechnology Research Efforts.* GAO/RCED-86-39BR. Washington, DC: U.S. Government Printing Office, 1985.

USDA-funded biotechnology research is conducted primarily in USDA's own research facilities and in state agricultural experiment stations or colleges of veterinary medicine that receive funding from USDA's Cooperative State Research Service. At the time of the study, USDA was funding, in whole or in part, 778 biotechnology research projects totalling $40.5 million.

Education for Farmers of the Future

*** 8.18 ***

United States. Department of Agriculture. *Research for Tomorrow.* The 1986 Yearbook of Agriculture, Washington, DC: U.S. Government Printing Office, 1986.

Another in the USDA's longstanding series of yearbooks, the 1986 volume describes and discusses current and prospective research programs in biotechnology and its applications to plants and animals. See the previous entry (* 8.17 *) for a brief enumeration of ongoing USDA projects directed to "research for tomorrow."

*** 8.19 ***

Hackett, Michael R., and James S. Long. "The Livestock Masters Program: It Works." *Journal of Extension* 23 (Fall 1985):11-13.

The Livestock Masters Program is a unique educational effort in the state of Washington in which adults who have raised livestock complete additional training and volunteer 50 hours during the year to advise other residents on small animal production. Two years experience and an evaluation suggest that the program participants are willing and able to offer credible information in small scale livestock production. The major value of the program is that volunteers are recognized as a link in the system to help others apply research-based knowledge to small scale enterprises.

*** 8.20 ***

Hurt, Chris, Vince Harrell, and Darden Kirby. "Are Your Farmers Confused about Marketing," *Journal of Extension* 24 (Winter 1986):15-17.

Marketing confuses and frustrates many farmers. The successful independent farmers of the present and future must be managers not only of production, but also of marketing and finance. The authors set up four broad goals in teaching marketing skills: change attitudes; encourage a plan for marketing integrated with production, financing, and risk-bearing; get marketing into a decision-making context; and help farmers monitor their progress.

The Farmer as Human Capital

*** 8.21 ***

Halcrow, Harold G. "Policy Analysis of Human Resource Input Markets." In *Agricultural Policy Analysis.* New York: McGraw-Hill, 1984. ISBN 0-07-025562-8.

"Human resources may be visualized as a form of capital, which has a supply and a demand function," Halcrow declares. Economic progress in agriculture requires two necessary and sufficient conditions: (1) a sufficient rate of investment in human capital to increase the productivity of human beings, and (2) a sufficient level of economic opportunity outside of agriculture to raise the cost of being in farming or other agricultural sectors. The first condition increases the demand for human capital in agriculture, and the second decreases and controls the supply.

*** 8.22 ***

Huffman, Wallace E. "Human Capital, Adaptive Ability, and the Distributional Implications of Agricultural Policy." *American Journal of Agricultural Economics* 67:2 (May 1985):429-434.

"A key question is the type of farm program that the United States should have in an unstable economic environment," Huffman points out. Farmers who have superior adaptive skills are expected on average to make better decisions. Furthermore, given the highly competitive nature of U.S. agriculture, successful adaptation to structural changes is selective. Farmers possessing poor adaptive skills can be expected to comprise a relatively large percentage of the persons forced by economic circumstances to seek alternative employment or retirement, provided governmental intervention does not neutralize the selection process.

*** 8.23 ***

Pingali, Prabhu L., and Gerald A. Carlson. "Human Capital, Adjustments in Subjective Probabilities, and the Demand for Pest Controls." *American Journal of Agricultural Economics* 67:4 (November 1985): 853-861.

In addition to risk aversion, farmer behavior in an uncertain environment is governed by subjective probability estimates of random events. Human capital development can improve farmers' ability to estimate pest damage probabilities. More accurate assessment of subjective probabilities leads to lower pesticide use and increases the use of labor-intensive pest controls. The human capital variables with the largest effects are formal schooling and farmer experience with smaller pesticide effects from field scouting and extension schools.

*** 8.24 ***

Rahm, Michael R., and Wallace E. Huffman. "The Adoption of Reduced Tillage: The Role of Human Capital and Other Variables." *American Journal of Agricultural Economics* 66:4 (November 1984): 405-413.

Rahm and Huffman present a model of adoption behavior and explain differences in farmers' decisions to adopt reduced tillage practices and the efficiency of farmers' adoption decisions. They show that the probability of adopting reduced tillage in corn enterprises differs widely across farms and depends on soil characteristics, cropping system, and size of farming operation. The results also show that farmers' schooling enhances the efficiency of the adoption decision.

*** 8.25 ***
Schultz, Theodore W. *Investing in People: The Economics of Population Quality*. Berkeley: University of California Press, 1981. ISBN 0520044371. LC 80-6062//r842.

In this book, Schultz stresses the overriding importance of investing in people and knowledge. Although he uses India as an example of a developing country in population quality, he also explores the increases in the value of human time in the United States, from 1900 to 1970.

*** 8.26 ***
Schultz, Theodore W., ed. *Distortions of Agricultural Incentives*. Bloomington: Indiana University Press, 1978. ISBN 0-253-31806-8. LC 78-3246.

Agricultural economists in countries with developing agriculture have concluded that education is more directly productive in farming areas with rapidly changing use of new technology. In traditional undeveloped agriculture, there is less evidence that education contributes to productive agriculture.

*** 8.27 ***
Sharples, D. Kent. "Rural Community Colleges: Venture Team for Success." In *New Dimensions in Rural Policy: Building upon Our Heritage*. S. Prt. 99-153. Joint Economic Committee, Congress of the United States. Washington, DC: U.S. Government Priting Office, 1986.

The community college has made higher education available at a lower cost in both rural and urban settings. As economic problems arose in the agricultural sector, the rural community college has assumed a natural leadership role in rejuvenating and restructuring its community's economic base. As a source for employment training and retraining, the community college performs a vital educational role.

Education for Farmers in Distress

*** 8.28 ***
Heffernan, William D., and Judith B. Heffernan. "Sociological Needs of Farmers Facing Severe Economic Problems." In *Increasing Understanding of Public Problems and Policies--1986*. Oak Brook, IL: Farm Foundation, 1987.

"Unfortunately, little attention has been focused on how the impacted families and communities can be assisted," the authors assert. The authors report in their study of farm families who had lost farms for financial reasons between 1980 and 1985, that the financial crisis is falling disproportionately on younger families. Those interviewed had been leaders in their communities; most had graduated from high school and had farms of various sizes and types. Nearly all had experienced depression and many had withdrawn from family and friends. Most had not received assistance from government agencies, churches, or other organizations.

*** 8.29 ***
Jones, Bruce L., and William D. Heffernan. "Educational and Social Programs as Responses to Farm Financial Stress." *Agricultural Finance Review* 47 (Special Issue 1987):148-155.

Educational and social programs are not widely understood because they have generally not been used or needed before. Six possible programs considered here are: information assistance; emotional support; legal assistance; financial counseling; vocational assistance; and financial assistance to exiting farmers. The need for farm financial assistance programs exists, in part, because of a lack of other programs to help farm families avoid severe stress when they go out of business. The need will continue as long as there are no educational and social programs that would at least moderate the hardships of financially stressed farm families. Social and educational programs designed to help farmers cope with financial and emotional problems are alternative farm financial assistance programs.

*** 8.30 ***
Weigel, Daniel J., Joan S. Blundall, and Randy R. Weigel. "Keeping Peace on the Farm: Stresses of Two-Generation Farm Families." *Journal of Extension* 24 (Summer 1986):4-6.

Some two-generation farm families find strength and support by working together; others find themselves unable to handle the emotional strains. Living and working in two-generation families can create special problems. With good communication, respect, and self-esteem, families can, much more effectively, handle the problems that occur.

State Efforts to Help Farmers

*** 8.31 ***
Gardner, Richard L. "The State Role in Addressing Farm Financial Problems." Paper presented at the annual American Agricultural Economics Association Meeting, Reno, Nevada, 27-30 July 1986.

A number of states have developed marketing and market development programs, motivated by concern for the survival of their farm economies, interest in improving the quality of food products available, and the search for opportunities to increase the value of farm products and create jobs in the process, before those products leave the state. The programs include development of cooperatives, direct marketing and promotional programs, and export marketing. State governments are working to diversify economic activity in rural communities and to maintain them as trade and residential centers. In addition, social and educational programs designed to help farmers cope with financial and emotional problems are sometimes alternatives to financial assistance programs.

Educational Programs for Those Who Leave Farming

*** 8.32 ***
Guither, Harold D., J. Paxton Marshall, and Paul W. Barkley. "Policies and Programs to Ease the Transition of Resources Out of Agriculture." *The Farm Credit Crisis, Policy Options and Consequences.* AE 4612. Urbana: Department of Agricultural Economics, College of Agriculture, University of Illinois, 1986.

As individual farmers and rural communities face the problem of excess resources in agriculture, they confront crucial decisions about their future. Individuals must decide if they can survive in farming, change careers, or undertake a skills training or schooling program. Agricultural communities must decide what options they have to stimulate economic development and employment opportunities for displaced farmers. The readjustment process underway will see many old and traditional actors leave the industry and many new and unknown actors enter. An estimated 30 million persons migrated from U.S. farms and rural areas between 1940 and 1980. During this period, farm numbers fell from 6.1 million to 2.4 million. One universal law relative to agriculture is that the agricultural labor force has to decline as development proceeds. When labor transfers out of agriculture, the per capita incomes of those who remain in rural areas are more likely to keep up with those in the non-farm sector.

*** 8.33 ***
Hildreth, R. J. "Policy Initiatives to Facilitate Adjustment." In *Increasing Understanding of Public Problems and Policies--1986.* Oak Brook, IL: Farm Foundation, 1987.

The transition to employment in the non-farm economy is neither easy nor automatic. Five broad policy issues should be considered in dealing with the situation. These are: assisting individual farmers and their families during their departure from farming; increasing area economic development; facilitating career reorientation; providing income support for displaced farmers during the adjustment period; and determining the role for government.

*** 8.34 ***
Mazie, Sara Mills, and Herman Bluestone. *Assistance to Displaced Farmers.* Agriculture Information Bulletin, no. 508. Washington, DC: Economic Research Service, U.S. Department of Agriculture, 1987.

"Leaving the farm causes terrible pain for many farm families, whether the decision is forced by a lender or made voluntarily by the family," Mazie and Bluestone conclude. Many farmers, unable to cope with heavy debts, have had to give up farming in the past few years. More are expected to leave agriculture over the next few years. Evolving federal, state, and local government programs, some described in this paper, are helping these farmers and their families to salvage as much equity and dignity as they can and to find other ways to make a living.

Challenges for the Future

*** 8.35 ***
Barkley, Paul W. "Using Policy to Increase Access to Capital Services in Rural Areas." In *Restructuring Policy for Agriculture: Some Alternatives.* Edited by Sandra S. Batie and J. Paxton Marshall. Blacksburg: College of Agriculture and Life Sciences, Virginia Polytechnic Institute and State University, 1984.

Major forces and trends in the general economy are limiting the access that rural areas have to various forms of capital. The author identifies human capital, infra-structural capital, and institutional capital as the vital forms of capital that determine the vitality of rural areas. He contends that rural areas and the people who live in them have been bypassed by policy at both the state and federal level.

*** 8.36 ***
Barnard, Freddie L. "Cooperation: A Key for Extension." *Journal of Extension* 13 (Summer 1985):8-9.

Barnard describes the joint effort of the Cooperative Extension Service and the Farmers Home Administration to implement the adoption of coordinated financial statements and the training of field staff and borrowers in their use. The training program would never have been more than an idea if it had not been for the cooperative attitude displayed throughout both organizations. Such a cooperative approach to adult education programs could become the norm rather than the exception.

*** 8.37 ***

Breimyer, Harold F. "Research and Education in Agriculture: Old Issues and New Biotechnology." *Economic and Marketing Information for Missouri Agriculture* 28:9 (September 1985).

Several kinds of change, or pressures for change, in agricultural research and extension have been underway for some time. Biotechnology provides a new dimension. The private versus public issue is scrambled by the many new kinds of research funds that have become available. Biotechnology goes far beyond everyday agricultural research. It involves a powerful set of tools including techniques for manipulation of DNA or the fusion of cells. This new technology offers enormous potential benefits and a new way of doing things. At the same time there is still a continuing need to be aware of the safety, ethical, and moral issues of genetic engineering.

*** 8.38 ***

Castle, Emery N. "Rural Institutions for the Future." In *New Dimensions in Rural Policy: Building upon Our Heritage*. S. Prt. 99-153. Joint Economic Committee, United States Congress. Washington, DC: U.S. Government Printing Office, 1986.

"It is ironic that a Nation that has been among the most progressive in the world in the development and settlement of its vast acreages should now find itself with major rural problems and with inadequate institutions to provide solutions," Castle points out. Many of the early rural institutions of this nation recognized that the rural community had a great deal of activity in addition to farming and agriculture. As times have changed, so, too, have the needs of rural residents. As agriculture and farming have become more sophisticated, the remainder of the rural community has become increasingly diverse. Castle calls for reforms that include modification of government programs, rural natural resource policies, flexibility for certain existing institutions, and a reaffirmation of the original broad mandate of the land grant institutions.

*** 8.39 ***

Comstock, Gary, ed. *Is There a Moral Obligation to Save the Family Farm?* Ames: Iowa State University Press, 1987.

Farmers, along with more than 25 authorities in economics, history, sociology, politics, ethics, and theology, contribute to this collection of 31 essays that examines the background and current state of the farm problem. It focuses on moral and ethical issues raised and moral and religious approaches to be considered in guiding America through difficult decisions concerning agricultural and food policy, food prices, corporate monopolies, federal budget deficits, and proper use of land. The current crisis is seen as part of a chronic problem, currently escalated to a critical state not seen since the Great Depression.

*** 8.40 ***

Douglass, Gordon K., ed. *Cultivating Agricultural Literacy*. Battle Creek, MI: W. K. Kellogg Foundation, 1985.

Few issues are of greater importance to the world than adequate food supplies, proper food use, and knowledge about the components of the agricultural industry. The editor of this volume concludes that many do not understand the complexities of America's food system nor its relationship to human nutrition or its impact on international relations and trade. The thesis of this book is that the liberal arts curriculum is inadequate to teach agriculture and that it would be significantly improved by augmenting its content with greater agricultural insight.

*** 8.41 ***

Eidman, Vernon R. "Restructuring Agricultural Economics Extension to Meet Changing Needs: Discussion." *American Journal of Agricultural Economics* 68:5 (December 1986):1310-1312.

Perhaps the most difficult dilemma extension faces is the staffing to offer quality educational programs on increasingly complex subject matter areas to an increasingly sophisticated clientele. An extension organization with more highly trained local specialists delivering educational programs and having direct linkages to academic departments has the potential for achieving higher rates of return for investments in applied research and extension educational programs.

*** 8.42 ***

Experiment Station Committee on Organization and Policy-Cooperative State Research Service, United States Department of Agriculture. *Research Perspectives*. Proceedings of the Symposium on the Research Agenda for the State Agricultural Experiment Stations. College Station: Texas Agricultural Experiment Station, 1985.

The symposium was designed to address the issues, concerns, opportunities, and methodologies to be considered in the development of a national research agenda for the State Agricultural Experiment Stations. Profitability is the key concern of the private sector and its representatives believe that the public sector must concern itself with keeping farmers in business. Productive efficiency and profitability must be a core focus of the research agenda of the State Agricultural Experiment Stations, and the maintainence of the United States as the world's leader in agriculture must also remain as a major focus.

*** 8.43 ***

Fox, Glenn. "Is the United States Really Underinvesting in Agricultural Research?" *American Journal of Agricultural Economics* 67:4 (November 1985): 806-812.

Although the idea that public investment in U.S. agricultural research is too low is widely accepted, Fox argues that there are two important limitations to the evidence: (a) many authors have compared the social rate of return to public investments with the private rate of return to private investments; and (b) costs of public expenditures on agricultural research have been underestimated by a failure to account for the marginal excess burden of the tax collection system. Taking these two factors into account, evidence in support of the underinvestment hypotheses is weakened considerably.

*** 8.44 ***

Knutson, Ronald D. "Restructuring Agricultural Economics Extension to Meet Changing Needs: Discussion." *American Journal of Agricultural Economics* 68:5 (December 1986):1298-1309.

Concerns for extension's future are rooted in past decisions regarding its mission and clientele, in the ever-increasing complexity of agriculture, in structural change within agriculture, and in organizational issues in the breakdown of the mission of the land grant system. To survive, Knutson calls for certain essential changes: clearly articulate the issues and present positive solutions; fully restore the tradition of extending research results; implement plans; address the issues related to organizational structure; implement expanded staff training and development programs; deal with the privatization issue; place increased emphasis on adapting and testing products of bio- and information technology; give special attention to the problems of moderate size farm survival; target educational programs; exercise leadership in interdisciplinary endeavors; keep programs timely; and accept reality of reduced federal formula funding.

*** 8.45 ***

Libby, Lawrence W. "Restructuring Agricultural Economics Extension to Meet Changing Needs: Discussion." *American Journal of Agricultural Economics* 68:5 (December 1986):1313-1315.

Extension and the rest of the land grant system may well be in trouble, but the mandate is to acknowledge and accommodate the reality of a diversifying rural America. Decisions on priority should be made by those in positions to deliver on them. Human capital investment involves both staying current and acquiring the initial stock of knowledge and experience.

*** 8.46 ***

Paarlberg, Don. "The Land Grant Colleges and the Structure Issues." *American Journal of Agricultural Economics* 63:1 (February 1981):129-134.

Paarlberg suggests that the land grant college system faces alternatives regarding the social consequences of its programs similar to those related to farm structure: "(a) continue to imply that technology is socially neutral; (b) continue the existing programs and frankly acknowledge that they have social consequences, good and bad; (c) modify the existing programs, taking into account their social consequences; or (d) a mixture of the three."

*** 8.47 ***

Peters, Robert R., Joe E. Manspeaker, and Estelle Russek-Cohen. "Bringing the Classroom to the Farm." *Journal of Extension* 24 (Spring 1986):8-10.

The authors point out that adults learn best when the facilities are easily accessible, provide an informal setting in which the audience feels free to participate, and allow for socialization. The farm fits all of these conditions. This approach is recommended for new extension workers and established workers who find declining attendance at in-town meetings.

*** 8.48 ***

Schuh, G. Edward. "Revitalizing Land Grant Universities." *Choices: The Magazine of Food, Farm,and Resource Issues* (Second Quarter 1986):6-10.

"The land grant universities have lost their way," Schuh declares. For the land grant universities to be relevant to the problems of society, they need major changes in their programs. For a variety of reasons many land grant universities find themselves paralyzed, Schuh charges. Administrators and faculty must re-instill a mission orientation. They must revitalize the tripartite mission of teaching, research, and extension. Everyone needs to recover a sense of institutional mission.

*** 8.49 ***

Van Horn, James E., Daryl K. Heasley, and Deborah Bray Preston. "Shedding the Cocoon of Status Quo." *Journal of Extension* 23 (Spring 1985):4-6.

The current pervasive social change is dramatically altering the family and the community in which extension workers have traditionally served. Such changes challenge educators to experiment with educational efforts to help these families despite risks of failure in breaking away from the traditional programs and methods.

*** 8.50 ***

Weber, Bruce A. "Extension's Roles in Economic Development." *Journal of Extension* 25 (Spring 1987):16-18.

The economic hard times of the early 1980s and the farm crisis place an urgency on how extension and other adult education institutions should be involved with economic development. Weber sees these roles for economic development education: providing perspective, increasing the knowledge base, teaching management skills, and shaping institutional structure. All state extension services need to examine the roles they play and decide which roles they can most appropriately play.

General Works

*** 8.51 ***

Rasmussen, Wayne E., and Gladys L. Baker. *The Department of Agriculture*. New York: Praeger Publishers, 1972. LC 77-117476.

"The future direction of the Department of Agriculture and its programs is in question," Rasmussen and Baker assert. Some supporters are calling for the department to concentrate on its traditional work with farmers; others urge that it emphasize consumers' interests in food and become what one secretary of agriculture advocated--a Department of food and agriculture. The issues which confronted the department in the early seventies are still at issue in the 1980s.

LANDMARKS IN AMERICAN AGRICULTURE

1607. The first permanent English colony in America was established at Jamestown, Virginia. See citations: 3.6

1620. The second permanent English colony was established at Plymouth, Massachusetts. Both of these colonies rejected the feudal farm land pattern common in Europe to establish family farms and a system of representative town government. See citations: 3.6

4 July 1776. The Declaration of Independence was proclaimed.

10 October 1780. The Continental Congress urged states to cede western lands to the nation, with the pledge that these lands would be settled and admitted as states.

1785. The Philadelphia Society for the Promotion of Agriculture was founded. It was an example of the new spirit of scientific improvement.

20 May 1785. The Ordinance of 1785 established a plan for disposing of western lands.

25 January 1786. In Shays' Rebellion, the farmers of western Massachusetts revolted against deflation and the financial policies of their Boston creditors.

13 July 1787. The Ordinance of 1787 established a system for governing western lands.

1793. Eli Whitney invented the cotton gin, which he patented in 1794. The machine led to cotton becoming the chief cash crop of the South and to slavery becoming more profitable.

14 March 1794. Eli Whitney received a patent on the cotton gin.

1796. The Land Act of 1796 authorized the sale of single sections of land.

7 December 1796. President George Washington recommended the creation of a national board of agriculture.

26 June 1797. Charles Newbold of New Jersey received a patent for the first cast-iron plow.

9 March 1820. The Land Law of 1820 was passed. Among its most prominent features were the discontinuation of the credit system and the reduction of the minimum price of lands from $2.00 to $1.25 an acre.

9 December 1825. The Committee on Agriculture and Forestry of the United States Senate was created.

1831. Cyrus H. McCormick invented his grain reaper in Virginia. The McCormick reaper reduced the workload of farmers harvesting grain, enabling them to expand their operation.

21 June 1834. McCormick was granted a patent on his grain reaper.

1836. Henry L. Ellsworth, commissioner of the newly created Patent Office, on his own initiative began to distribute seeds and plants by means of the franking privilege of cooperating Congressmen.

1837. John Deere began manufacturing plows in Illinois with steel share and smooth wrought iron moldboard.

28 June 1837. A practical threshing machine was patented by the Pitts Brothers.

3 March 1839. Congress appropriated $1,000 to the Patent Office for collecting agricultural statistics, conducting agricultural investigations, and distributing seeds.

1841. The Preemption Act, providing for the sale of previously settled public lands at $1.25 an acre to the original settlers, was passed.

3 April 1848. The Chicago Board of Trade, the oldest futures market in the United States, was organized to accommodate a rapidly expanding cash grain crop.

12 February 1855. Michigan passed legislation providing for the establishment of the Michigan Agricultural College.

23 February 1855. Pennsylvania Farmers' High School, later renamed Pennsylvania State College and now named Pennsylvania State University, was established by the state legislature.

15 May 1862. President Abraham Lincoln signed the legislation which created the U.S. Department of Agriculture. See citations: 3.1, 3.6, 6.1, 8.51

20 May 1862. The Homestead Act was approved by President Lincoln. See citations: 3.1, 3.6, 6.1

1 July 1862. Isaac Newton of Pennsylvania served as first commissioner of agriculture from 1 July 1862 to 19 June 1867.

2 July 1862. President Lincoln approved the Morrill Land-Grant College Act. See citations: 3.1, 3.6, 6.1, 8.12, 8.13, 8.46, 8.48

25 December 1865. Chicago's Union Stockyards opened.

4 December 1867. Oliver Hudson Kelley, an employee of the U.S. Department of Agriculture, organized the Patrons of Husbandry, later known as the National Grange. This was the first general farmers organization to admit women to equality of membership and privilege.

3 March 1873. The Timber Culture Act allowed papers to be taken out on 160 acres of treeless land. The homesteaders agreed to plant a quarter of their land with trees over a 10-year period.

29 May 1884. The Bureau of Animal Industry was established in accordance with an act of Congress.

2 March 1887. The Hatch Experiment Station Act was approved, providing federal grants to states for agricultural experimentation.

9 February 1889. The U.S. Department of Agriculture was raised to cabinet status.

30 August 1890. The second Morrill Land-Grant College Act authorized separate land grant colleges for Negroes; 17 were established. See citations: 3.1, 3.6, 6.1

1892. The first successful gasoline tractor was built by John Froehlich.

21 October 1895. Sunkist Growers, Inc., for many years called the California Fruit Growers Exchange, was incorporated as the Southern California Fruit Exchange.

28 August 1902. The Farmers Union, also known as the Farmers Educational and Cooperative Union of America, was organized.

30 June 1906. The Pure Food and Drug Act was approved.

30 June 1906. The Meat Inspection Act was approved.

10 August 1908. President Theodore Roosevelt organized the Country Life Commission, which made its first report to him on 23 January 1909.

1911. The first Farm Bureau was formed in Broome County, New York. See citation: 6.3

16 May 1913. The Office of Markets, the first marketing agency, was established in the U.S. Department of Agriculture. This marked the formal beginning of organized marketing programs in the federal government. See citations: 4.37,-4.59, 4.60, 4.61, 4.62, 4.63

8 May 1914. The Smith Lever Act formalized cooperative agricultural extension work. See citations: 8.3, 8.4, 8.36, 8.44

1916. The National Milk Producers' Federation was formed; in 1923, "Cooperative" was added to its title.

17 July 1916. The Federal Farm Loan Act, providing for 12 farm land banks, was approved. The law grew out of Country Life Commission recommendations.

16 August 1916. The U.S. Grain Standards Act authorized official standards for grain and required inspection of grain sold by grade in interstate or foreign commerce. See citations: 7.47, 7.48, 7.49, 7.50, 7.51

23 February 1917. The Smith-Hughes Vocational Education Act was approved. See citation: 8.4

10 August 1917. President Wilson established the Food Administration by executive order.

10 August 1917. The Food Production Act was approved.

13 November 1917. The Food Administration announced hog price supports fixed at a level in relation to corn, but this price was not maintained.

1918. The development of a system for growing modern hybrid seed corn was completed by Donald F. Jones.

18 June 1918. A presidential proclamation required stockyards and livestock dealers to be licensed. This was the first federal regulation of livestock market practices.

1 July 1918. Sugar rationing went into effect. The Sugar Equalization Board was incorporated to allocate and distribute sugar supplies; to a large extent, it supplanted the international Sugar Committee.

3 March 1920. The American Farm Bureau Federation was formally organized and its constitution ratified. See citation: 6.3

1921. The Farm Block was organized in the Congress to support agricultural legislation independent of party lines.

15 August 1921. The Packers and Stockyards Act authorized the Secretary of Agriculture to regulate meatpackers and livestock trading practices at public markets having an area of 20,000 square feet or more.

24 August 1921. The Grain Futures Trading Act was passed. It was invalidated by the Supreme Court, was revised, and became the Grain Futures Act of 1922. See citations: 4.41, 4.42

18 February 1922. The Capper-Volstead Act declared that a cooperative association was not, by reason of the manner in which it was organized and normally operated, a combination in restraint of trade in violation of the federal antitrust statutes. See citations: 4.21, 4.22, 4.23, 4.24, 4.25, 4.26, 7.60

21 September 1922. The Grain Futures Act was approved. See citations: 4.41, 4.42

21 September 1922. The Commodity Exchange Act was passed to regulate trading in certain commodity exchanges.

16 January 1924. The first McNary-Haugen farm relief bill was introduced into Congress. See citation: 6.4

24 February 1925. The Purnell Act, providing funds for economic and sociological research carried on by state experiment stations, was passed.

1 February 1926. The domestic allotment plan was proposed to raise farm prices and income. See citation: 6.9

2 July 1926. Congress passed the Cooperative Marketing Act, which created a Division of Cooperative Marketing in the U.S. Department of Agriculture.

25 May 1928. The McNary-Haugen bill was vetoed for the second time by President Coolidge. See citation: 6.4

15 June 1929. The Agricultural Marketing Act became law. The Federal Farm Board was established. See citations: 3.17, 4.37

May 1932. The Farm Holiday Movement was led by Milo Reno.

4 March 1933. Henry A. Wallace of Iowa served as secretary of agriculture from 4 March 1933 to 4 September 1940.

12 May 1933. The Emergency Farm Mortgage Act was approved.

12 May 1933. The Agricultural Adjustment Act was approved. The Agricultural Adjustment Administration established. See citations: 6.1, 6.3, 6.7, 6.8, 6.10, 6.12, 6.34, 6.71, 7.8, 7.11, 7.18, 7.21

27 May 1933. The Farm Credit Administration was established as an independent agency. It became part of the U.S. Department of Agriculture.

16 June 1933. The Farm Credit Act, providing for the reorganization of agricultural credit activities, was passed.

19 September 1933. The Soil Erosion Service, which later became the Soil Conservation Service, was created in the U.S. Department of the Interior. See citations: 2.24, 7.69, 7.73, 7.76, 7.77, 7.78, 7.80, 7.81, 7.88

4 October 1933. The Farm Credit Administration requested the governor of each state to establish a committee to attempt the conciliation of excessive and distressed farm debts.

17 October 1933. The Commodity Credit Corporation was established. See citations: 6.1, 6.3, 6.7, 6.8, 6.10, 6.12, 7.8, 7.11

11 May 1934. Great dust storms originated in the "Dust Bowl" area of the Great Plains region. The drought was the worst ever recorded in the United States; it extended over 75 percent of the country and severely affected 27 states.

12 June 1934. The Reciprocal Trade Agreements Act was approved.

28 June 1934. The Frazier-Lemke Farm Bankruptcy Act was approved.

28 June 1934. The Taylor Grazing Act gave the U.S. Department of the Interior power to regulate grazing on public domain in the West. See citation: 7.70

27 April 1935. Congress declared soil erosion a national menace in an act directing the U.S. Department of Agriculture to establish a Soil Conservation Service. See citations: 2.24, 7.69, 7.73, 7.76, 7.77, 7.78, 7.80, 7.81, 7.88

11 May 1935. The Rural Electrification Administration was established by Executive Order 7037 and was incorporated into the U.S. Department of Agriculture on 1 June 1939.

6 January 1936. The Agricultural Adjustment Act was invalidated by the Hoosac Mills decision of the U.S. Supreme Court.

29 February 1936. Congress passed the Soil Conservation and Domestic Allotment Act as a substitute measure for the Agricultural Adjustment Act.

11 May 1936. The Rural Electrification Act was approved. Activities authorized by the act previously had been conducted under the Emergency Relief Appropriation Act of 1935.

15 June 1936. The Commodity Exchange Act (formerly the Grain Futures Act) was approved.

16 February 1938. The Agricultural Adjustment Act of 1938 provided for farm price support and adjustment programs based upon an "ever-normal granary" concept. It replaced and invalidated the Agricultural Adjustment Act of 1933.

16 February 1938. The Federal Crop Insurance Corporation was created as an agency within the U.S. Department of Agriculture by Title V of the Agricultural Adjustment Act of 1938.

13 March 1939. The Food Stamp Program was formally announced as an experimental program by Secretary of Agriculture Henry Wallace and the chairman of the National Food and Grocery Conference Committee.

28 May 1940. A school penny-milk program under Section 32 of amendments to the Agricultural Adjustment Act was approved by the secretary of agriculture.

26 May 1941. Congress raised price supports for major agricultural commodities to 85 percent of parity through loans on the crops.

1 July 1941. The Steagall Amendment provided for price supports for expansion of production of nonbasic agricultural commodities.

4 August 1942. The Migrant Labor Agreement with Mexico became effective.

15 September 1942. By the fall of 1942, it became necessary to institute rationing of farm machinery, which was delegated to the U.S. Department of Agriculture by the Office of Price Administration. The program was ended on 21 November 1944.

13 November 1942. The U.S. Congress passed the Tydings Amendment to the Selective Training and Service Act, which was a "remain working in agriculture or fight" clause. The amendment was intended to stop farmers from leaving for higher paying jobs elsewhere.

5 December 1942. The War Food Administration powers were transferred to the U.S. Department of Agriculture by Executive Order 9280. The agricultural press hailed the establishment of a "Food Administration" within the Department of Agriculture, although this name was not officially used until four months later under Executive Order 9334.

4 June 1946. The National School Lunch Act, which authorized assistance to states through grants-in-aid and other means in establishing nonprofit school lunch programs, was approved.

14 August 1946. The Farmers' Home Administration Act was passed and approved as Public Law 731. The act abolished the Farm Security Administration.

30 October 1947. The General Agreement on Tariffs and Trade was signed.

1 January 1948. The General Agreement on Tariffs and Trade became effective. See citations: 4.47, 4.51

3 July 1948. The Agricultural Act of 1948 was approved.

31 December 1948. The U.S. Department of Agriculture's obligation under the Steagall amendment to support specified nonbasic commodities at 90 percent of parity was terminated.

13 June 1949. The International Wheat Agreement was approved by the Senate.

31 October 1949. The Agricultural Act of 1949 incorporated the principle of flexible price support and provided a change in the parity formula. It also provided, through Section 416, for additional domestic disposition of surplus agricultural commodities for donations to needy persons abroad through U.S. voluntary relief organizations.

1954. The National Wool Act supported prices to encourage domestic production.

10 July 1954. The Agricultural Trade Development and Assistance Act of 1954 (Public Law 480) revised Section 416 of the Agricultural Act of 1949 to encourage export of price-supported commodities to nations unable to make purchases on the world market and to aid agricultural improvement in developing nations.

28 August 1954. The Agricultural Act of 1954 established flexible price supports, authorized commodity set-asides, and provided wool support payments.

28 August 1954. The Special School Milk Program was established under the Agricultural Act of 1954, which provided authority for use of Commodity Credit Corporation funds to increase fluid milk use in schools.

1 September 1954. The Social Security Act was amended to extend coverage to farm operators.

1955. The National Farmers Organization was established.

1956. The Rural Areas Development Program was initiated by the U.S. Department of Agriculture on a pilot, or demonstration, basis in a few counties.

28 May 1956. The Agricultural Act of 1956 (Soil Bank) included provisions for federal financial assistance to farmers for converting general cropland into conservation uses.

28 August 1958. The Agricultural Act of 1958, providing for more effective prices, production adjustments, and marketing programs for various agricultural commodities, became law.

21 September 1959. Legislation was approved authorizing the secretary of agriculture to carry out a food stamp program.

22 March 1961. The Feed Grain Act was approved.

3 April 1961. The Economic Research Service was established in the U.S. Department of Agriculture. The functions transferred from the Agricultural

Marketing Service and the Agricultural Research Service had been assigned to the Bureau of Agricultural Economics prior to the reorganization of 1953.

May 1961. The president authorized inauguration of an experimental Food Stamp Program and established eight pilot projects.

8 August 1961. The Agricultural Act of 1961 was approved. It established programs for the 1962 wheat and feed grain crops, authorized marketing orders for several farm commodities and the Special Milk Program, and extended Public Law 480.

21 May 1963. Wheat producers rejected a mandatory acreage control plan. See citation: 6.24

11 April 1964. The Agricultural Act of 1964 was approved, establishing voluntary cotton and wheat programs.

31 August 1964. The president approved a nationwide Food Stamp Act. See citations: 6.43, 7.7

3 November 1965. The Food and Agriculture Act of 1965 was approved with the goal of encouraging farmers to adjust production between various crops.

30 November 1970. The Agricultural Act was passed initiating a cropland set-aside program for producers of wheat, feed grains, and upland cotton.

11 January 1971. The Food Stamp Act of 1964 was amended to make the program available to more people. See citations: 6.43, 7.7

10 August 1973. The Agriculture and Consumer Protection Act was approved.

29 September 1977. The Food and Agriculture Act of 1977 was passed, which provided price and income protection for farmers and assured an abundance of food and fiber at reasonable prices to consumers. See citation: 1.10

14 December 1977. A new farm group, the American Agriculture Movement, called for a national strike by farmers for better prices. Members rallied to support the strike with a parade of farmers, livestock, and tractors.

31 January 1979. Secretary of Agriculture Bob Bergland declared the United States free of hog cholera after 99 years of research and 17 years of a federal-state eradication campaign.

5 February 1979. An American Agriculture Movement tractorcade of farmers, most of whom were grain producers, arrived in Washington to lobby Congress and the administration for higher prices.

12 March 1979. Secretary of Agriculture Bob Bergland proposed a full-scale national dialogue on the future of American agriculture with an emphasis on developing a workable policy on the structure of U.S. agriculture. See citations: 1.22, 1.23, 3.12

22 December 1981. President Reagan signed the Agriculture and Food Act of 1981, the culmination of another evolutionary state in public policymaking. See citations: 6.26, 8.16

23 December 1985. President Reagan signed the Food Security Act of 1985, a part of the continuing evolution of U.S. agricultural and food policy. Major changes included setting loan rates based on market prices and more emphasis on soil conservation. See citations: 6.48, 6.49, 6.50, 6.51, 6.52, 6.53, 6.54, 6.55, 6.56, 6.57, 6.58, 6.59, 6.60, 6.61, 6.62, 6.63, 6.64, 6.65, 6.66, 6.67, 6.68, 7.15, 8.1

AGRICULTURAL AND FOOD POLICY TERMS

The vocabulary of agriculture and agricultural policy includes many commonly used words but adds new meanings to them. Our glossary of terms is designed to help clarify the special meanings of terms used in farming and the food and fiber system. The list includes many new terms emerging from the passage of the Food Security Act of 1985. Probably no list can ever be complete because new words and phrases always seem to be entering the agricultural vocabulary.

ACRE - The basic unit of land measurement containing exactly 43,560 square feet. A section of land one mile square contains 640 acres.

ACREAGE ALLOTMENT - An individual farm's share, based on its previous production, of the national acreage needed to produce sufficient supplies of a particular crop; currently used only for tobacco. See also: FARM ACREAGE and CROP ACREAGE BASE

ACREAGE LIMITATION PROGRAM - See: ACREAGE REDUCTION PROGRAM.

ACREAGE REDUCTION PROGRAM (ARP) - A program that requires a farmer to reduce the amount of crop planted below his base acreage to qualify for price supports and target prices for that crop, if such a program is in effect for that crop.

ADVANCE PAYMENTS - Payments made in advance of the time that data are available to compute the exact program benefits.

AGRIBUSINESS - Firms engaged in production and distribution of agricultural inputs or in the marketing, processing, or distribution of agricultural commodities.

AGRICULTURAL ADJUSTMENT--A term generally referring to programs designed to regulate or control agricultural production and marketing. The Agricultural Adjustment Act of 1933 created the Agricultural Adjustment Administration in the Department of Agriculture. Since then adjustment programs have been implemented through similar agencies under various names.

AGRICULTURAL CONSERVATION PROGRAM (ACP) - A program in which producers agree to carry out specified conservation practices on their farms and receive payments to help pay part of the cost.

AGRICULTURAL POLICY - A broad term used to encompass those government programs that most directly affect the prices and incomes received by farmers.

AGRICULTURAL STABILIZATION AND CONSERVATION SERVICE (ASCS) - The agency in the U.S. Department of Agriculture that administers the farm price and income support programs as well as some conservation and forestry cost sharing programs. In addition to the office in Washington, DC, there are offices in each state and most counties.

ALLOTMENT - An allotted share of production for an individual farm based on previous production. Allotments are currently applicable only to peanuts and tobacco price support programs.

ASCS- See: AGRICULTURAL STABILIZATION AND CONSERVATION SERVICE.

ASSETS - Cash and other property that have a market value.

BALANCE OF TRADE - The difference between the amount of exports and imports. The balance is positive if exports exceed imports or negative when imports exceed exports.

BALANCE SHEET OF AGRICULTURE - An accounting statement showing the total value of land and other property that farmers own, the amount of debt that farmers owe, and the difference between the two, usually called the net worth.

BASE ACREAGE - See: FARM ACREAGE BASE and CROP ACREAGE BASE.

BASIC COMMODITIES - In the price support legislation of the 1930s and 1940s, six agricultural crops were designated as basic commodities--corn, cotton, peanuts, rice, tobacco, and wheat--and specific price support programs were designated for these crops.

BASIC LOAN RATE - In the Food Security Act of 1985, the basic loan rate is that rate set by law. In 1986, the basic loan rate for corn is $2.40 and for wheat $3.00 per bushel. In 1987 and later years, the basic loan rate cannot drop more than 5 percent below the previous year. The secretary of agriculture has authority to reduce the actual loan rate below the basic rate by as much as 20 percent. See also: FINDLEY PROVISION.

BEEF PROMOTION AND RESEARCH ACT - Part of Title XVI of the Food Security Act of 1985 that provides for a check-off and promotion program for beef.

BLENDED CREDIT - A financing plan for export sales in which government credit guarantees, or government credit at lower interest rates, is blended with regular commercial credit to provide lower interest rates and more favorable terms for foreign buyers.

BUFFER STOCKS - Supplies of a product stored on farms or in commercial elevators or warehouses to moderate extreme price fluctuations by assuring a more stable supply. Buffer stocks are usually controlled by government while total stocks include both government and privately held stocks.

CAPITAL GAINS - When property is sold for more than the owner paid for it, the difference between the purchase price and the sale price, after all expenses are paid, is called the capital gain.

CARGO PREFERENCE - A policy requiring that a certain portion of goods or commodities exported from the United States be shipped in American ships.

CARRYOVER - The supply or volume of a farm commodity not yet used at the end of a marketing year. It continues to be stored or is used during the following marketing year. Carryover may be referred to as the end of year stocks.

CARTEL - An alliance or arrangement among enterprises in the same field of business aimed at securing an international monopoly. A cartel usually seeks to control production or the amount marketed to raise prices and maximize profits.

CASH FLOW - The total funds generated by a farm or firm for covering costs and investment. Farming presents unique cash flow problems when income is generated only when crops are harvested or livestock is ready for market, if crop failures occur or costs rise faster than product prices.

CCC- See: COMMODITY CREDIT CORPORATION.

CEREAL GRAINS - Grains used for human food such as wheat, rice, or rye.

CHECK-OFF - A program by which a small amount of money, per-unit of product, is deducted from the proceeds of a farm commodity when it is sold, usually by the first buyer, for the purpose of supporting research or promoting sales of that product.

COALITION - A combination of organizations and groups working together to influence a single piece of legislation or government decision.

COARSE GRAIN - See: FEED GRAINS.

COMMODITIES - Broadly defined, any goods exchanged in trade, but usually used to refer to widely traded raw materials and agricultural products such as wheat or corn.

COMMODITIES FUTURES TRADING COMMISSION (CFTC) - The agency of the federal government responsible for regulating and overseeing the operations of all futures contract markets.

COMMODITY CREDIT CORPORATION (CCC) - A government owned and operated corporation authorized to borrow funds from the U.S. Treasury to operate the U.S. Department of Agriculture's price and income support programs, to manage government owned stocks of agricultural commodities and administer their disposal through domestic and export programs. Most activities are carried out by ASCS personnel, although certain programs are administered and implemented through the Agricultural Marketing Service, the Foreign Agricultural Service, and the Food and Nutrition Service.

COMMODITY CREDIT CORPORATION SALES PRICE - The CCC may not sell its stocks of wheat, corn, sorghum, barley, oats, rye, or cotton at less than 115 percent, or its stocks of rice at less than 105 percent of the current national average loan rate of each commodity. If loan repayments are permitted at lower than loan rate levels, the resale price is 115 percent and 105 percent of the average loan repayment rates for these crops.

COMMODITY PROGRAMS - A collective term used to include the price support programs for corn and other feed grains, wheat, cotton, rice, peanuts, tobacco, sugar, and dairy products.

COMMON AGRICULTURAL POLICY (CAP) - The agricultural policy of the European Economic Community.

COMMON MARKET - See: EUROPEAN ECONOMIC COMMUNITY.

COMPARATIVE ADVANTAGE - The situation when a farm or a country produces and sells those goods and services, which it can produce at relatively low cost, and buys those products and services, which others can produce at relatively less cost, the central concept in modern trade theory.

CONSERVATION RESERVE - A program where highly erodible land is retired from crop production and planted to grass or trees for a period of years and for which the owner receives an annual payment. See also: HIGHLY ERODIBLE CROPLAND.

CONSERVATION USE - An approved use of land that protects soil from erosion by its being planted to grasses, legumes, or to small grains that are not allowed to mature.

CONSIDERED PLANTED - In calculating base acreage, land is considered planted to a program crop if it was used for conservation acreage under an acreage reduction or set-aside program; if the producer was prevented from planting the crop because of drought, flood, other natural disaster, or other condition beyond the control of the producer; or acreage planted to a nonprogram crop if planted on land that was permitted to be planted to the program crop; and any acreage on the farm that the secretary of agriculture determines is necessary to be included in establishing a fair and equitable crop acreage base.

CONSUMER SUBSIDY EQUIVALENTS - The level of subsidy that would be necessary to compensate consumers for the removal of government programs.

CONTRACT SANCTITY- The legal doctrine that the terms of a contract are inviolable.

COOPERATIVE - A form of business owned by the customers. There are many types of farmer-owned cooperatives that provide supplies and services or buy and sell agricultural commodities. The Capper-Volstead Act of 1922 exempted cooperatives from restrictions of the antitrust law and established government support and assistance to farmer cooperatives as a national policy.

CORPORATE FARM - A farm business that is legally incorporated under state law. The stock may be held by a farm family, closely held and not available for public purchase, or it may be listed on a public stock exchange. The term may be used incorrectly when referring to large farm operations, which are in fact sole proprietorships or partnerships.

COST-OF-PRODUCTION - The average amount, in dollars-per-unit, to grow or raise a farm product, including all purchased inputs and sometimes including allowances for management and use of land owned by the farm operator. The cost may be expressed in units of a bushel, pound, ton, or per-acre, depending on the product involved.

COST SHARING - In certain farm conservation programs, the government will share the cost of certain farm practices with the farm owner or operator.

COVER CROP - A close growing crop, such as a grass or legume, grown primarily to protect and improve soil between periods of regular crops, or in orchards and vineyards.

CROP ACREAGE BASE - As defined in the Food Security Act of 1985, the crop acreage base is the average number of acres planted and considered planted to a program crop for harvest in the previous five years. Where the program crop was not grown on the farm in each of the previous five years, the law provides for procedures to develop a crop acreage base for that farm.

CROP INSURANCE - A program operated or insured by the Federal Crop Insurance Corporation in which farmers can purchase insurance against crop disasters. The producers pay about 70 percent of the cost and the government pays about 30 percent of the total program costs.

CROSS COMPLIANCE - The requirement that a farmer who wishes to participate in a price support program and qualify for price support and loans in that program must also meet the program provisions for other major program crops that he or she grows. See also: OFF-SETTING COMPLIANCE.

DAIRY PRICE SUPPORT PROGRAM - The program established by Congress in the Agricultural Act of 1949. From 1949 until 1981, milk prices were supported by law at 75 to 90 percent of parity. Since 1981, Congress has passed several bills that have set lower minimum support prices for milk. Prices are supported by government purchases of manufactured dairy products to maintain the minimum price for milk established by Congress.

DEBT/ASSET RATIO - A measure used to determine the financial soundness of a farming operation. Farmers whose debts are equal to 70 percent or more of their assets are considered to be in some financial difficulty.

DEFICIENCY PAYMENTS - Federal government funds paid to farmers when farm prices are below the target price. The payment rate is determined by taking the difference between the target price and the average price received by farmers for the marketing year, or the average price for the first five months of the marketing year, or the loan rate depending upon the commodity, and the determinations of the secretary of agriculture. The total payment is determined by multiplying the payment rate, times the acreage base, times the program payment yield.

DISASTER PAYMENTS - Federal government payments to farmers participating in certain government programs when the farmers are prevented from planting their crop or when their crop yields are abnormally low. Under the 1985 farm bill, the Secretary of Agriculture is required to make such payments in areas where federal crop insurance is not available. He may also make such payments if crop insurance is inadequate or when he determines that such payments are necessary to avert an economic emergency.

DIVERSION PAYMENT - Payments made to farm owners and operators for diverting land from certain crops into conservation uses.

DIVERTED ACRES - Under acreage reduction programs, those acres that are taken out of production and diverted to some conserving use.

DOUBLE CROP - In some areas a second crop can be planted after the major crop is harvested. For example, soybeans may be grown after the wheat crop is harvested. Government farm programs may be designed to permit this normal cropping practice in certain parts of the country.

ECONOMIC EMERGENCY DISASTER PAYMENTS Producers of wheat, feed grains, upland cotton, and rice may be eligible for an economic emergency disaster payment if a disaster has reduced production and that loss of production results in an economic emergency.

ELASTIC DEMAND - A market situation in which the percentage change in price will bring about a greater proportional change in the amount purchased so total receipts will be greater with a lower price.

ELIGIBLE PRODUCER - A producer who is eligible for certain government farm program benefits by signing an agreement to carry out certain practices.

EMERGENCY LOANS - Loans made to farm producers under emergency credit programs, usually for conditions resulting from drought, floods, or other natural disasters.

EQUITY - The net worth of an individual farmer

or business firm, the net value of property after all debts are deducted.

ESTABLISHED PRICE - See: TARGET PRICE.

EUROPEAN ECONOMIC COMMUNITY (EEC) - A federation of twelve European countries organized to promote economic growth and trade between the member countries. The countries are: Belgium, Denmark, France, Greece, Great Britain, Ireland, Italy, Luxembourg, Netherlands, Spain, Portugal, and West Germany. Spain and Portugal joined the EEC in 1986.

EXPORT CERTIFICATES - A discretionary provision in the Food Security Act of 1985 by which the secretary could make export certificates available to producers participating in the wheat and feed grain programs. The certificates would be redeemable for cash when the certificate holder can show that the amount of grain shown on the certificate has been exported.

EXPORT ENHANCEMENT PROGRAM - Various programs in which the government pays subsidies to increase the volume of agricultural exports.

EXPORT PIK - An export subsidy program by which the government provides exporters with special bonuses in the form of commodities so they can compete in the international market. Export PIK commodities were to be given only to countries in which competing exporting countries were also using export subsidies. See also: PAYMENT-IN-KIND.

EXPORTS - The goods, services, and products which are sold to buyers in foreign countries.

EXPORT SALES REPORTING - The USDA policy requires that export sales in one day involving more than 100,000 metric tons of major grains and oilseeds be reported to USDA within 24 hours of sale. For other commodities, weekly reports are required.

EXPORT SUBSIDY - A government grant, made to a private enterprise, for the purpose of facilitating exports. In Europe, it is often termed "restitution."

FAMILY FARM - A farm in which a family provides most of the labor, management decisions, and operating capital. The land may be owned, partly owned, or rented. Some economists estimate that most family farms would have annual sales between $40,000 and $300,000.

FARM - Starting in 1978, the Bureau of the Census defined a farm as any place that has or would have had $1,000 in gross sales of farm products.

FARM ACREAGE BASE - The total of crop acreage bases on the farm. In 1987 and later years, the farm acreage base will include all crop acreage bases plus the average acres planted to soybeans in 1986 and later years and the average acreage devoted to conservation uses in 1986 and later years.

FARM CREDIT ADMINISTRATION - An independent federal agency that supervises the farm credit system, operating through 12 districts each of which comprises several states.

FARM CREDIT SYSTEM - The credit institutions established by authority of Congress and that are now farmer owned, the federal land banks, the federal intermediate credit banks, production credit associations, and banks for cooperatives. The federal government still supervises the system through the Farm Credit Administration.

FARM MANAGEMENT - The science and art of combining land, labor, and capital to establish and operate farming operations efficiently and profitably.

FARM MARKETING QUOTA - See: MARKETING QUOTA.

FARM PROGRAM PAYMENT YIELD - See: PROGRAM YIELD.

FARMER-OWNED RESERVE (FOR) - A program of long-term Commodity Credit Corporation loans for wheat and feed grains intended to stabilize prices and hold reserves for times of short production. Under the program, farmers who place their grain in storage receive an extended nonrecourse loan for 3 to 5 years. Interest on the loan may be waived and farmers may receive annual storage payments from the government. The farmer cannot take his grain out of storage without penalty until the market prices reaches a specific "release price." At that point, the farmer may decide to remove his grain from the reserve but is not required to do so. In an emergency, the USDA can require repayment of the loans. The program was first established in the 1977 Food and Agriculture Act and was extended with modifications in 1984 and 1985.

FARMERS HOME ADMINISTRATION (FmHA)- The agency in the U.S. Department of Agriculture that is authorized to make direct loans to farmers who cannot obtain credit from other sources. It may also guarantee to farmers loans made by banks. The agency also makes loans for rural housing and community facilities as authorized by Congress.

FEDERAL CROP INSURANCE - A voluntary program available to farmers since the 1930s. The program was revised under legislation passed by Congress in 1980. At present, farmers pay about 70 percent of the cost and government pays about 30 percent.

FEDERAL LAND BANK - See: FARM CREDIT SYSTEM.

FEDERAL MARKETING ORDERS AND AGREEMENTS - See: MARKETING ORDERS AND AGREEMENTS.

FEDERAL RESERVE BOARD - See: MONETARY POLICY.

FEED GRAINS - Those grains most commonly used for livestock or poultry feed. Corn, grain sorghum, oats, and barley are the main feed grains produced in the United States. Sometimes they are called coarse grains.

FINDLEY PROVISION - A provision in the Agriculture and Food Act of 1981 and carried over into the Food Security Act of 1985, which gives the secretary of agriculture discretion to reduce crop loan rates below the basic loan rate by up to 20 percent if the average market price in the previous marketing year was not more than 110 percent of the loan rate for that year, or if the reduction is necessary to maintain a competitive market position.

FOOD GRAINS - The cereal grains most commonly used for human food. Wheat, rice, and rye are the main food grains produced in the United States.

FOOD SECURITY ACT OF 1985 - The comprehensive farm bill that replaced the Agriculture and Food Act of 1981 and sets U.S. farm policy through 30 September 1990.

FOOD SECURITY IMPROVEMENTS ACT OF 1986 The bill passed in March 1986 to make technical corrections and adjustments in the Food Security Act of 1985. Major items included rules for establishing program payment yields, underplanting of permitted acres and use of nonprogram crops, targeted export assistance, haying and grazing on diverted acreage, and increase in dairy assessments.

FREE MARKET - A market in which prices are set by competitive forces without direct government influence.

FREE TRADE - A theoretical concept to describe international trade unhampered by governmental barriers such as tariffs, quotas, and embargoes.

FUTURES CONTRACT - A contract to buy or sell a set amount of a commodity for delivery at a future time and place. The two largest futures trading markets in the United States are the Chicago Board of Trade and the Chicago Mercantile Exchange.

GENERAL ACCOUNTING OFFICE (GAO) - An agency of Congress that investigates the operations of various programs and the expenditure of appropriated funds. Most investigations are conducted at the request of Congressional committees or individual members of Congress.

GENERAL AGREEMENT ON TARIFFS AND TRADE (GATT) - Multilateral agreement signed by 92 countries, which establishes rules and guidelines for regulating world trade among members and a forum for countries to discuss and resolve trade disputes. An underlying principle of the GATT is that trade should be restricted only through the use of uniformaly applied tariffs.

GENETIC ENGINEERING - A highly technical process of developing new plants and animals by combining genes into new forms and combinations. The process is also called biotechnology.

GRAMM-RUDMAN-HOLLINGS - An act passed in 1985 that specified rules for moving toward a balanced budget through systematic cuts in federal expenditures. Provisions of the Gramm-Rudman-Hollings Act could lead to sharply reduced farm program expenditures as long as the federal budget deficit remains.

GUARANTEED LOAN - An arrangement by which the government guarantees repayment of a loan made by a private lender. The Farmers Home Administration may guarantee some loans made to farmers by private banks. The government may guarantee loans made to exporters by private banks to promote more agricultural exports.

HIGHLY ERODIBLE CROPLAND - As defined in the Food Security Act of 1985, land that is in crop land use and is classified by the Soil Conservation Service as class IV, VI, VII, or VIII, under the land capability classification system in effect at the time the law was enacted, is highly erodible.

HOMESTEAD PROTECTION - A provision in the Food Security Act of 1985 that permits a foreclosed borrower to remain in the principal residence on the farm even though the land is repossessed.

IMPORTS - The goods, services, and products that a country buys from other countries.

INDEMNITY PROGRAMS - Payments made under certain programs to producers who sustain losses as a result of pesticides, nuclear radiation, fallout, residues, or toxic substances.

INELASTIC DEMAND - A market situation in which a change in price will bring about a smaller proportional change in the amount purchased. For example, consumers tend to buy and consume about the same amount of a product regardless of the change in price, with the result that less total expenditures are made for a large output than for a small one. An elastic demand has the opposite characteristics.

INELASTIC SUPPLY - Supply is considered inelastic if the increase in production is relatively less than the change in price.

INNOVATORS - Those persons who are the first to develop or try new methods or practices.

INPUT - Items used in the production of an agricultural product such as seed, fertilizer, chemicals, feed, machinery, fuel, labor, and land.

INTEGRATION - The combining of various steps in the production and marketing of a product under the management or control of a single firm. See also: VERTICAL INTEGRATION.

INTEREST BUY-DOWN - A plan to assist persons in financial difficulty with large debts in which the lender may reduce the interest rate and the government pays part of the interest cost so total interest payments are reduced.

INTERNATIONAL TRADE BARRIERS - Regulations used by governments to restrict imports from, and exports to, other countries. Examples are tariffs, embargoes, import quotas, and unnecessary sanitary restrictions.

LEVEL PLAYING FIELD - The concept that farmers should have an equal opportunity to compete for markets or have equal treatment under government policies.

LIQUIDITY - The ability to supply funds or raise money by selling assets to pay off debts.

LOAN DEFICIENCY PAYMENT - The difference between the loan repayment rate and the rate at which the loan was obtained when a marketing loan is in effect. This term may also apply to a situation whenever the secretary lowers the formula-determined loan rate up to an additional 20 percent for wheat and feed grains as he did in 1986. In this case, payments must be made to producers to provide the same return as if the loan rate had not been reduced.

LOAN FORFEITURE - The forfeiting of commodities placed under loan instead of repaying the loan in cash.

LOAN RATE - The price per unit (bushel, bale, pound) at which the government will provide loans to farmers to enable them to hold their crops for later sale. See also: NON RECOURSE LOANS.

LOBBYISTS - Individuals and organizations that attempt to influence decisions by members of Congress and other government officials.

MACROECONOMIC POLICIES - Policies that affect the general economic environment in which the total economy or sectors, such as agriculture, operate. Monetary and fiscal policies are examples.

MANDATORY PROGRAM - A program that would require all farmers to participate.

MARKET-ORIENTED FARM POLICY - A policy with an objective of letting prices be set primarily in a public market rather than by government actions.

MARKET PRICE - The amount received or paid for a unit of a commodity. Most market prices of farm commodities are quoted in bushels, pounds or hundred pounds, tons, or dozens.

MARKETING CERTIFICATES - Certificates issued by the Department of Agriculture as part of the price support program that may be redeemed for cash or commodities, depending upon the specific provisions of the program.

MARKETING LOAN - A loan that may be repaid at a level below the rate at which the loan was made.

MARKETING ORDERS AND AGREEMENTS - Federal regulatory programs that permit agricultural producers collectively to promote orderly marketings of a crop or commodity. Orders and agreements are requested by producers and issued by the secretary of agriculture. To go into effect, an agreement or order must be approved by producers in a referendum, usually by a two-thirds majority. The marketing order permits producers to join together to regulate marketing of the commodity by means of restrictions binding on all handlers of the commodity in the area covered by the order. Such restrictions may involve packing standards, grades, size, price, and limitations on quantities shipped or marketed. Agreements may be issued in conjunction with orders or may be issued without orders. If issued without orders, there are less restrictions on marketing procedures and only apply to those producers or handlers who agree voluntarily to enter into the agreement with the secretary of agriculture.

MARKETING QUOTA - The quantity of a crop that will provide adequate and normal market supplies at the national level. This quantity is translated into acreage or individual farm marketing quotas based on a farm's previous production of that commodity. In the past, for marketing quotas to be in effect, producers had to approve by a two-thirds majority in a referendum. When in effect, a producer cannot market more than his quota without penalty. For certain tobaccos, a poundage quota may apply as well as an acreage allotment.

MARKETING SPREAD - The difference between the price the producer received for a commodity and the price paid by the consumer for an equivalent amount of the same product.

MARKETING YEAR - The period of 12 months beginning at the start of harvest of a crop and extending to the same time in the following year. The marketing year for wheat, oats, and barley begins on June 1 and extends through May 30 the following year. For corn, grain sorghum, and soybeans starting with 1986, the marketing year is from September 1 through August 31.

MICROCOMPUTER - Small computers that fit on a desktop and can be used for many business, accounting, and financial operations.

MILK ASSESSMENTS - Deductions from dairy farmer milk checks to finance special federal government supply reduction programs. See: MILK PRODUCTION TERMINATION PROGRAM.

MILK DIVERSION PROGRAM - The program passed by Congress in 1983 by which milk producers could receive direct payments by agreeing to reduce production in their herds between 1 January 1984 and 31 March 1985.

MILK MARKETING ORDERS - The programs covering much of the fluid milk marketed in the United States that fix minimum prices that buyers must pay.

MILK PRODUCTION TERMINATION PROGRAM (Whole-Herd Buyout) - A program established by the Food Security Act of 1985 under which the secretary of agriculture paid dairy farmers to quit producing milk for a period of five years.

MONETARY POLICY - Policies carried out by the Federal Reserve Board to influence the supply of money and the rate of interest.

MORATORIUM - A postponement in payment of interest and principal on debts owed by farmers, particularly those with financial problems. The proposals for a moratorium on farm debts are usually aimed at those debts farmers owe to the U.S. government through loans from the Farmers Home Administration.

MULTILATERAL AGREEMENT - An international compact in which three or more parties participate.

NATIONAL FARM PROGRAM ACREAGE - The number of harvested acres of a crop that is estimated to be needed to meet domestic and export use and to accomplish any desired increase or decrease in carryover.

NATIONAL WEIGHTED AVERAGE MARKET PRICE - The average price received by producers for a commodity taking into consideration the amounts marketed at different prices and at different locations.

NEGOTIABLE MARKETING CERTIFICATES - Marketing certificates that can be exchanged for cash or commodities. See: MARKETING CERTIFICATES.

NET WORTH - The value of property a person owns after subtracting all the debts that he or she owes.

NOMINAL RATE OF INTEREST - The actual rate of interest paid without any adjustment for inflation.

NO NET COST PROGRAMS - Price support programs in which producers are assessed to finance the cost of the program. The tobacco program was designed to be a "no net cost" program to the federal government.

NON RECOURSE LOANS - Price support loans to farmers to enable them to hold their crops for later sale. The loans are nonrecourse because if a farmer cannot profitably sell the commodity and repay the loan with interest when it matures, the commodity for which the loan was advanced can be delivered to the government in full settlement of the loan. The farmer can redeem his commodities by paying off the loan with interest.

NON-TARIFF TRADE BARRIER - Any type of restraint on imports or exports other than a tariff.

NORMAL CROP ACREAGE - The acreage on a farm normally devoted to a group of designated crops.

NORMAL YIELD - A term designating the average historical yield established for a particular farm or area. Normal production would be the normal acreage planted to a commodity multiplied by the normal yield.

OFF-SETTING COMPLIANCE - The requirement that a farmer who wishes to participate in a program for one farm must also meet the program provisions for other farms that he or she owns or operates. See also: CROSS-COMPLIANCE.

OILSEED CROPS - Those crops from which oil is extracted from the seed--primarily soybeans, peanuts, cottonseed, and flaxseed. Sunflower, safflower, castor beans, and sesame are also produced for oil and are considered oilseed crops.

ORGANIC FARMING - Farming methods that use only organic fertilizers and avoid use of inorganic agricultural chemicals and herbicides.

PAID DIVERSION PROGRAM - A program that provides direct payments to farmers in return for diverting a specified amount of acreage of certain crops into conserving use.

PARITY - A relationship that defines a level of purchasing power for farmers equal to an earlier base period. The base period as defined by law and used in calculating parity prices is 1910-14. Some farmers, rather than use the technical definition above, think of parity as simply "a fair price plus a reasonable profit."

PARITY INDEX - The index of prices paid by farmers for items used in production, interest, taxes, and wage rates. Each month the U.S. Department of Agriculture issues the current index along with the index of prices received by farmers, parity prices on individual commodities, and the parity ratio.

PARITY PRICE - The price for a commodity that would give it the same purchasing power that it had in the base period.

PARITY RATIO - The ratio of the prices received by farmers to prices paid, based on the indexes issued monthly by the U.S. Department of Agriculture.

PAYMENT-IN-KIND (PIK) - A program that provides payment to farmers in the form of commodities for reducing acreage of certain crops and placing that acreage in conserving uses. The term may also apply to export enhancement programs or other programs where payments are made in the form of commodities.

PAYMENT LIMITATION - A limit set by law on the amount of money any individual farmer may receive in farm program payments each year under the wheat, feed grains, cotton, and rice programs. The limit under the 1985 Agriculture and Food Act is $50,000; $100,000 is the limit for disaster programs.

PERMANENT LEGISLATION - The laws upon which many agricultural programs are based. For the major commodities, the permanent legislation is the Agricultural Adjustment Act of 1938 and the Agricultural Act of 1949. These laws have been frequently amended for a given number of years, but would again be in effect if current amendments are not enacted. If Congress had not passed the Food Security Act of 1985, the Agricultural Act of 1949 and Agricultural Adjustment of Act of 1938 would have remained in effect.

PIK - See: PAYMENT-IN-KIND.

PIPELINE STOCKS - The minimum quantity of any product needed to carry on the normal processing and marketing operations.

POLICY - A course of action. Public policy refers to actions taken by a government body.

PORK PROMOTION, RESEARCH, AND CONSUMER INFORMATION ACT - Part of the Food Security Act of 1985 that provided for a check-off, referendum, and a producer-operated promotion and research program for pork.

PREVENTED PLANTING - A situation in which a producer is unable to plant the crop in a field that he intends to because of wet weather at planting time, floods, or other natural disaster.

PRICE SUPPORTS - Government programs that aim to keep farm prices from falling below specific minimum prices. Most price support programs are carried out by providing loans to farmers so they can store their crops during periods of low prices, making direct purchases of certain commodities, or making direct payments under certain conditions.

PRODUCER ASSESSMENTS - Assessments charged against producers of a commodity to help pay the cost of a specific price support program. Assessments are being used in the tobacco and dairy price support programs in 1986.

PRODUCER SUBSIDY EQUIVALENTS - The level of subsidy that would be necessary to compensate producers, in terms of income, for the removal of government programs affecting a particular commodity.

PRODUCTION CONTROL PROGRAMS - Any government program intended to limit production. At various times these programs have been called acreage reduction, reduced acreage, set-aside, diverted acreage, acreage allotments, marketing quotas, PIK, and soil bank.

PRODUCTION CREDIT ASSOCIATION - See: FARM CREDIT SYSTEM.

PRODUCTIVITY - The quality of being productive.

PROGRAM BENEFITS - The various forms of financial assistance available to those farmers who sign up and agree to comply with the requirements of government farm programs. These benefits may include eligibility for Commodity Credit Corporation loans, deficiency payments, diversion payments if offered, and disaster payments under certain circumstances.

PROGRAM YIELD - The yield for a crop on a given farm used to calculate deficiency payments. Program yields are based on history of past yields, and records of crop sales provided to the local ASCS office by the individual producer.

PROVEN YIELD - Yields substantiated by records of crop sales or other documentation acceptable to local ASCS offices.

RANGELAND - Land, usually in the West, that is used for grazing of animals rather than for growing crops.

REAL RATE OF INTEREST - The rate of interest earned after deducting the average rate of inflation.

RECOURSE LOAN - A loan in which commodities could be used as collateral, but which would have to be repaid in cash rather than delivering the commodities, which is permitted under a non recourse loan.

REDUCED ACREAGE PROGRAM - See: ACREAGE REDUCTION PROGRAM.

REFERENDUM - In relation to agricultural policy, a referendum involves a vote by producers of a specific commodity for a proposed program that will obligate all producers to participate in the program if a specified percentage of producers favor it.

REGRESSIVE - A policy or program that has a greater effect on, or works to the disadvantage of, lower income persons.

RELEASE PRICE - The price at which farmers who have grain stored in the reserve may sell it without incurring penalties. The release price for one corn reserve was $3.25 per bushel in 1984.

RESERVE PREMIUMS - In some instances, producers entering their grain into the reserve have been eligible to add reserve premiums to their loan rates. Farmers who placed 1982 wheat in the reserve received a reserve premium of 45 cents a bushel and for 1982 corn, 35 cents a bushel.

SANCTITY OF CONTRACTS - See: CONTRACT SANCTITY.

SET-ASIDE - A program under which a farmer must divert a proportion of his crop land to soil conserving uses to be eligible for program benefits.

SIZE NEUTRAL - A policy or program that would affect all farms, or give benefits, in such an amount to neither encourage nor discourage change in size of farm operation.

SOD BUSTER BILLS - Bills introduced in Congress designed to prevent the plowing up of rangeland and the planting of grain crops. The Food Security Act of 1985 had sodbuster provisions that restricted plowing up of highly erodible land for crop production and required approved soil conservation plans on highly erodible land in crop production in order to be eligible for government farm program benefits.

SOIL BANK - A program operated in the 1950s to achieve both soil conservation and production control objectives. Under the program, farmers signed contracts for varying periods of time to place part of their acreage into conserving uses.

STOCKS - See: CARRYOVER.

STORAGE PAYMENT - The payment a farmer receives from the government when he places a commodity in the farmer-owned reserve and stores it in his own storage facilities or in commercial storage facilities.

STRUCTURE OF AGRICULTURE - The make-up of the agricultural sector--usually described in terms of numbers and sizes of farms, types of farm business organizations, and other features that determine the control of agricultural assets and management decisions.

SUBSIDY - A government payment. Subsidies are paid to many people and firms under many different programs.

SUPPLEMENTARY PAYMENTS - A payment similar to deficiency payments made to producers of wools, mohair, and extra-long staple cotton. Wool and mohair producers may receive payments equal to the percentage difference between the support price and the national average market price times their annual revenue from the sale of wool and mohair.

SUPPLY MANAGEMENT - A term used to describe a policy in which government programs are used to influence and control the supply of a commodity to maintain a desired price.

SWAMPBUSTER BILLS - Legislation that places restrictions on the draining of natural wetlands for crop production.

TARGETED EXPORT ASSISTANCE - Programs designed to increase exports to specific countries by means of various types of subsidies.

TARGETING - See: TIERING.

TARGET PRICE - A price for certain crops established by law. If the average market price does not equal the target price, qualifying farmers receive a deficiency payment to make up the difference. Generally deficiency, or target price payments are made if average market prices do not equal the target price for the first five months of the marketing year. The target price for 1986 corn is $3.03 per bushel. See: DEFICIENCY PAYMENTS.

TAX LOSS FARMING - Farming operations carried out with the main objective to produce a loss for tax reporting purposes. Such losses from farming operations may be used to reduce taxes owed from other businesses or nonfarm income sources.

TECHNOLOGY - In agriculture, the methods, techniques, and systems used in farm production and marketing of farm products.

TENANCY - The renting or leasing of land by a farm operator.

TENANT PROTECTION - Provisions in some farm legislation are designed so that both tenants and landowners are entitled to program benefits.

TENURE - The relationship or type of control that a farmer has on the land that he farms--usually described as owner, part-owner, or tenant.

TIERING - A method of directing benefits of federal price support programs toward smaller or medium size farms. Under tiering, deficiency payments per unit of production would be higher for a limited number of bushels and lower for quantities beyond that amount.

TRIGGER PRICE - The point at which the market price reaches or exceeds the release price of grain in the reserve for the specified length of time, so farmers can sell their reserve grain without penalty.

TWO-PRICE PLAN - A plan that involves supporting that part of production used in the domestic market at one price and selling the remainder for export at whatever it will bring.

VERTICAL INTEGRATION - A situation where two or more firms at different stages of production and processing-marketing combine under a single ownership or management.

WHOLE-HERD BUYOUT - See: MILK PRODUCTION TERMINATION PROGRAM.

Author Index

Title Index

Titles enclosed in quotation marks refer to either articles in journals or parts of composite books. Titles without quotation marks refer to separately published books.

In descending order of the frequency with which they have been cited, the principal journals relied on in this bibliography are: *American Journal of Agricultural Economics, North Central Journal of Agricultural Economics, Choices: The Magazine of Food, Farm, and Resource Issues*, and *Agricultural Finance Review.*